The Brain as a Tool

The Brain as a Tool
A Neuroscientist's Account

Ray Guillery

OXFORD
UNIVERSITY PRESS

OXFORD
UNIVERSITY PRESS

Great Clarendon Street, Oxford, OX2 6DP,
United Kingdom

Oxford University Press is a department of the University of Oxford.
It furthers the University's objective of excellence in research, scholarship,
and education by publishing worldwide. Oxford is a registered trade mark of
Oxford University Press in the UK and in certain other countries

First Edition published in 2017
Impression: 1

Published in the United States of America by Oxford University Press
198 Madison Avenue, New York, NY 10016, United States of America

British Library Cataloguing in Publication Data
Data available

Library of Congress Control Number: 2017935885

ISBN 978-0-19-880673-8

Printed and bound by
CPI Group (UK) Ltd, Croydon, CR0 4YY

Oxford University Press makes no representation, express or implied, that the
drug dosages in this book are correct. Readers must therefore always check
the product information and clinical procedures with the most up-to-date
published product information and data sheets provided by the manufacturers
and the most recent codes of conduct and safety regulations. The authors and
the publishers do not accept responsibility or legal liability for any errors in the
text or for the misuse or misapplication of material in this work. Except where
otherwise stated, drug dosages and recommendations are for the non-pregnant
adult who is not breast-feeding

Links to third party websites are provided by Oxford in good faith and
for information only. Oxford disclaims any responsibility for the materials
contained in any third party website referenced in this work.

This book is dedicated to Murray Sherman, with thanks for many critical and helpful contributions to my thoughts about the brain and for many years of friendship and cooperation.

Preface

We rely on our brains for all aspects of our conscious lives, in health and disease. Neuroscientists need to understand the ways in which nerve cells communicate with each other and how the messages that are passed from one nerve cell to another link us to the world. Advances in understanding how messages pass along nerve fibres and from one cell to the next are well advanced in terms of understanding these events at molecular levels. Our understanding of what the messages mean and how the neural activity relates to our cognitive and behavioural lives is, in contrast, much more limited.

That is the focus of this book, and the title is an attempt to see how we depend on using the brain, its nerve cells and neural pathways, to learn about our interactions with the world. The brain on its own can do nothing. We use our brains to interact with the world. It is not until we start moving about, moving our eyes, or our fingers, that we can start using our brains to learn about the world by interacting with it. If we want to understand animals including ourselves and each other in health and disease, we need to understand the neural pathways and their functions. Models or theoretical proposals are most useful once they can be related to identifiable brain parts that can be studied down to molecular levels. That is the focus of this book: to raise answerable questions about living brains and identifiable intercommunicating nerve cells.

This book follows three earlier books on the pathways to the cerebral cortex (Sherman and Guillery 2001, 2006, 2013) written jointly with S Murray Sherman, now at the University of Chicago. It owes much to those books and to my more than four decades of lively, thoughtful, and productive interactions with Murray. I now see that the implications of the neural results summarized in those books lead to important conclusions, not only about how we plan and interpret further experiments, but also beyond that, to the territories of psychologists and philosophers concerning the brain's highest functions. This book explores where, in order to understand these functions, knowledge about specific neural pathways and their actions will be relevant.

In the past, there have been two distinct views about how our brains relate to our bodies and the world: the standard view of most current neuroscience textbooks has the cerebral cortex receiving sensory messages from the world and then reacting to what it has received by sending messages out to the muscles. This, the sensory-to-motor view, is a realist view based on a real world that we can describe in our daily lives or read about in physics textbooks. An alternative view, an idealist or interactive view, has so far received more support from psychologists and philosophers than it has from neuroscientists. This view recognizes that we use the brain to *learn* about the meanings of the messages that the brain receives from the body and the world. Whereas the sensory physiologist identifies these meanings easily, knowing the details of stimuli, the cortex receives only the message, a pattern of neural impulses comparable to a Morse code. For an organism like us to understand or respond appropriately to these messages we have to use the brain as a tool for exploring the relevant parts of the world, moving our fingers to identify objects in the dark, moving our eyes to see the objects in the light, or moving our whiskers if we are a mouse. These sensorimotor interactions depend on the fact that each incoming message actually has two meanings, one about the sensory event and the other about the *forthcoming* action that is already being processed at lower levels of the brain.

A brief account of both views in Part I of the book shows how the standard view was established mainly between the 1850s and the 1950s by tracing sensory and motor pathways through the brain. Part I also introduces a number of pathways that have played a lesser role, or no role at all in the textbooks, but that can today be seen to provide strong neural evidence for the interactive view. I show throughout the later parts of the book that the questions we can ask about the brain, and the experiments we can get funded, depend significantly on which of these two views our questions are based on. The difference matters not only to philosophers and psychologists, it matters particularly to neuroscientists.

In the first chapters of Part II, I go back to the early 1950s, when I was learning to be a neuroanatomist (neuroscientists were unheard of at the time). I do this partly to introduce a historical background and a personal angle to the book, but mainly to illustrate how a subject that had been growing rapidly since the mid-nineteenth century, describing the major pathways of the brain in terms of their connections and their functions, was suddenly lacking what had seemed until the 1950s to have been its own firm theoretical basis. Studies of nervous systems relied heavily on the 'neuron doctrine' together with a 'law of dynamic polarization' about the organization of individual nerve cells. Neither the doctrine nor the law survived undamaged through the 1960s and I was fortunate to have been taught that neuroanatomy is not a part of a separate discipline with its own doctrines and laws, but a part of biology, based on contemporary views of evolutionary relationships, where animals must be studied in terms of the uses that the parts serve in the lives of the animals. The brain has an evolutionary history and its parts can be expected to have specific uses. We still need to learn how we and our vertebrate relatives use our brains. In the later chapters of Part II, I apply such an evolutionary view of the brain to a cell group called the thalamus, an obligatory relay on the way to the cerebral cortex. Since this is present in mammals but largely lacking in our vertebrate ancestors, we face a key question. What is it about the thalamus and cortex that distinguishes mammalian behavioural and cognitive capacities from those of our ancestors? I was unable to answer this then, during the 1960s, but the question led me to study details of the structures in the thalamus, and these gradually led me to the contents of this book.

Part III is heavily based on my earlier publications with Murray Sherman between the 1970s and 2013. It describes the thalamic relay of the visual pathways to the cerebral cortex and shows that the thalamus as a whole serves as a relay and a gate for all the messages that reach the cerebral cortex, some coming from other sensory pathways, some from other lower centres of the brain, and some coming from the cerebral cortex itself. These last pathways, the transthalamic corticocortical pathways, revealed a feature that has proved to be common to all of the pathways relayed in the thalamus of mammals. The messages passing through the thalamus to the cerebral cortex travel along branched axons, one branch carrying information about events in the world, the body, or the brain to the thalamus for relay to the cortex, and the other carrying motor instructions on their way to lower centres that control the muscles. Since a branched axon sends essentially the same message down both branches, the cortex receives a copy of motor instructions that are on their way to execution elsewhere. The transthalamic corticocortical connections are arranged in a hierarchical order with several levels, each area serving to monitor lower levels and also to contribute to the relevant motor controls at lower levels as needed.

Part IV explores the cognitive and behavioural implications of this hierarchical arrangement of cortical monitors. I describe each cortical area as receiving a dual message, one about a recent event in the world, the body, or the brain, and another about an instruction for a forthcoming related action, which in its turn will produce a new sensory message and a further, slightly later action. This generates the flow of sensorimotor interactions that provides a continuity to the flow of our conscious lives. Often this flow is interrupted by an unexpected event, perhaps in the world, perhaps in the brain. Then the mechanisms of the cortical hierarchies and the functions of the thalamic relay take over. As higher cortical levels monitor the lower levels, they can identify the problem and send their own motor instructions to correct the error; often with a copy of the instructions sent to a yet higher cortical area through a thalamic relay. In this way, the continuous flow of perceptions that lead to actions, which, in turn, lead to new perceptions, can produce and maintain the continuous flow of our conscious lives, providing us with a sense of the continuity of ourselves and the world. We describe them as conscious actions because we can anticipate them, thus clearly distinguishing them from the action of others.

When we experience a novel, unexpected sensory input, perhaps while learning a new skill, perhaps learning to back a car into a parking area or viewing the world through inverting lenses, the smooth flow of sensorimotor actions is not immediately achieved. The cortical hierarchy must be brought into play to control our phylogenetically older motor centres, guiding them at difficult, unexpected moments until a well-integrated sensorimotor sequence is once more created. We can start to think of the cortex as a tool that is particularly useful for helping us to learn new skills.

If we ask why mammals in particular may have acquired this new hierarchy of cortical monitors, the answer may well be that mammals give birth to living young, commonly several, that are breastfed as a group. The new, early challenges of the social interactions required by such a family may well have required the rapid modifications of sensorimotor interactions that the cortical hierarchy of monitors can produce. Once the new mechanism had been established, their potential for new learning was, and almost certainly still is, almost unlimited.

For whom has this book been written? At first for neuroscientists, especially for young neuroscientists interested in the functional organization of the neural pathways, who are looking for new ways of seeing the brain and for new questions to ask about it. I also hope that it will interest many older neuroscientists and people in related fields, not only in psychology and philosophy but also in the many related fields that are contributing to our understanding of the brain, people who are looking to understand the functions of the brain from the point of view of their own disciplines. I have included a set of illustrations that should serve as a guide to the parts of the brain discussed for those who have a limited background in neuroscience. There is also a glossary of terms used. Both of these are designed specifically for the items that play a role in the book; they should not be regarded as a simplified primer in neuroscience generally.

There is an important message in the illustrations. They will strike a neuroscientist as representing the past rather than the future. That is deliberate. I am describing and reinterpreting the past, in order to raise issues that need to be studied in the future. There are now many new methods for studying the structures and functions of nerve cells that have enormous promise, but have not yet produced new conceptual structures for understanding the functional relationship of the brain's pathways. The illustrations in this book relate to the conceptual structures we have, and the text raises issues about where new evidence can in the future be found, probably by one or another of the new methods, methods which are still increasing. The book is an effort to make sense of the information we have inherited from the past. It is written for those who will be generating new conceptual structures in the future, and who will be able to illustrate them with some of the wonderful images that the new methods generate.

This book owes thanks to many people. Outstanding among these, as the contents of the book demonstrate, are two people: one was JZ Young, my teacher, thesis supervisor, and senior colleague from 1948 to 1964 in the Department of Anatomy at University College London. He taught me much and gave me the freedom to learn from my mistakes; the other is Murray Sherman, who has shared my interest in the puzzles of the brain since the late 1960s, and has contributed to many joint publications and three books that provided a basis for the present book. Even more, he has been a good friend and valuable colleague for many years. Our conversations, discussions, and arguments often left the brain for other subjects, but always tended to come back to the same subject: the brain and our ignorance as to its functions.

Many others have contributed to my thoughts about the brain, and I will only list some of them here. I am grateful to them all. Lizzie Burns has earned my thanks not only for the figures themselves, but also for her enthusiasm about preparing them, and her interest in the text itself. Colin Beesely has helped with many of the copies of figures from the literature. Others who have been involved in the preparation of this book in one of many different ways include Andrew Parker, Andy Smart, Anna Mitchell, Beycan Gödze Ayhan, Carol Mason, Emily Mackevicius, Filiz Onat, Fritz Sommer, Husniye Haçiolu Bey, Kouichi Nakamura, Kutay Deniz Atabay, Paul Bolam, Peter HJ Ralston, Richard Boyd, Sowon Park, Sten Grillner, Stuart Judge, Charlotte Holloway, and April Peake. My thanks to them all.

Ray Guillery

Contents

Part I

How do we relate to the world?

Chapter 1

The role of the brain

1.1 Summary

This chapter introduces two interpretations of how we know about the world. One, the standard, sensory-to-motor view, is that physical actions for sounds, lights, tastes, smells, and so on act on our sense organs to produce messages that are sent through the nervous system to the **cerebral cortex**, where the relevant structures of the world can be recognized and appropriate motor actions can be initiated. The other is an interactive sensorimotor view where the nervous system receives information about our interactions with the world, abstracting our knowledge about the world from these interactions. These two opposing views have rarely been considered in terms of specific neural pathways or the messages that they carry; that is the plan for this book, which will show that each view leads to different sets of interpretations of experiments and to different sets of research proposals.

The final part of the chapter explores a well-studied and widely taught clinical condition that illustrates the confusions that can arise when the dual, sensorimotor meaning of the incoming messages is not recognized.

1.2 Knowing the world

The puzzle of just how we relate to the world around us is one that has long concerned philosophers, psychologists, and occasionally also neuroscientists. How do we know the world? Do we know the objects of the world—the apples, trees, people, houses—directly as primary objects that we can identify on the basis of the smells, tastes, touches, lights, sounds, temperatures, or pressures that impinge upon us from the world? This is the standard or sensory-to-motor view of the world. Or do the sensory inputs provide the primary and only information we have about the world, providing **perceptions** that need further interactions before we can learn about the world itself? This is the interactive sensorimotor view. Do we *first know* the world, or our perceptions? The relevance of this difference for an experimental neuroscientist forms a significant part of the argument of this book.

We can think of the former as a realist view, based on an unquestionable real world that is present to our senses, and the latter as an idealist view, based on our individual and private perceptions of the world. We may decide that from a practical point of view for our daily lives we don't have to make the distinction. Most of us manage quite well, treating the world as made up of real objects, also recognizing that the perceptions on which we base our knowledge of the world are entirely private mental events, called qualia or first-person experiences by philosophers. When we see a red apple, it seems pretty straightforward that this apple is red. However, if we then see an apple in a painting, we may begin to see that the colour of apples is more complicated; there begin to be questions about how we relate our perceptions to the world, about what the world is 'really' like. In the rest of the book, I shall argue that we can best find answers to these questions by understanding the nervous system itself.

Hume (1739/2010) considered that our 'impressions' and 'ideas' are our only direct source of knowledge about the world (see also Hume 1748/1999). He wrote:

> there is a direct and total opposition betwixt our reason and our senses; or more properly speaking, betwixt those conclusions we form from cause and effect, and those that persuade us of the continued and independent existence of body. When we reason from cause and effect, we conclude, that neither colour, sound, taste, nor smell have a continued and independent existence. When we exclude these sensible qualities there remains nothing in the universe, which has such an existence. (Hume 1739/2010, final paragraph of section IV)

Helmholtz, who made significant contributions to physics, neuroanatomy, and sensory physiology, not only recognized the importance of examining the border between natural and mental science, but also, on the basis of his studies of the **retina** (Helmholtz 1868/1971, page 148), wrote: 'The question whether it is possible to maintain the natural and innate conviction that the quality of our sensations of sight give us a true impression of the qualities of the outer world' is 'that they do not'. He used the term 'unbewusster Schluss' to describe how we relate our perceptions to the world. The German term used by Helmholtz is usually translated as an unconscious inference about what our senses are telling us about the world, or even as a hypothesis. It seems more realistic to speak of unconscious *conclusions*, translating the 'Schluss' as an end of the matter with no hesitation or doubt implied.[1] We don't go through life making inferences or hypotheses about the world. We have arrived at, and we act on some very definite *conclusions* about what the world is really like. Our unconscious conclusions about a real world are firmly held and work extremely well.[2] We have to learn early in life from our interactions with the world that certain *sensorimotor* relationships can lead to extraordinarily firm conclusions about what the world is 'really' like. These unconscious conclusions derive from, and are learnt on the basis of, our *interactions* with the world. They probably represent some of the infant's earliest conscious experiences of interactions with the world. They serve in our daily lives as we interact with parts of what appears to us as adults to be a real world, and it is puzzling to see them challenged.

In order to understand how we arrive at these unconscious conclusions about the world, which serve as the shifting focus, 'the theatre', of our conscious lives, I will look at how our nervous systems relate to the world, how they transmit messages that are about the world, and how they equip us to interact successfully with the world. We need to learn how our brains serve in our interactions with the world, and we will find, in agreement with Helmholtz and Hume, that there is no reason to believe in a real world that we can identify directly through our senses, and no reason to see this as a worrying or puzzling conclusion.[3] Rather,

---

[1] When as children we went on talking after my mother's goodnight she would come back and tell us, 'mach Schluss'.

[2] The process of arriving at the conclusions should be regarded as unconscious even when the conclusion is clearly a conscious experience. See Chapters 12 and 13.

[3] Varela et al. (1993, page 133) have written, 'Why do we become nervous when we call into question that there is some way that the world is "out there," independent of our cognition, ...?'.

it is something to be understood as following necessarily from the way in which our brains interact with the world. The challenge of defining the neural processes that generate our conscious perceptions of the world has not yet been met. We need to define the neural circuits most likely to be involved and to understand their actions as an individual learns about novel sensorimotor interactions.

The first three parts of this book are concerned with details of neural structures and functions that can produce these successful interactions. As we define the neural links on which we can base a new view of our interactions with the world, many difficult problems about perception can be seen in a new light. Some of these problems are explored in the final parts of the book. Whereas in the past the problems have often been recognized as well-established puzzles about perception, supporting one or another non-neural interpretation of perceptual processing, my aim is to see the puzzles in terms of neural functions. I will first suggest a new way of understanding the functions of the neural pathways and then consider how this can address some well-recognized problems concerning perceptions and actions in relation to the world.

The account of the neural pathways is the primary goal of this book and has to be seen as a work in progress that draws attention to many details of neural connections and neural functions that become newly relevant in a new context when one turns from the standard textbook view of sensory-to-motor relationships to an interactive sensorimotor view. The neural pathways, and the ways we know them, form an important part of the book because they point to new research projects, needed to challenge and expand our knowledge of the relevant neural connections. One of my aims is to look for clues that will lead to worthwhile questions and stimulate new investigations of the structures and functions of the brain. In the past few years, neuroscience has acquired a number of remarkably powerful new techniques that allow detailed studies of individually labelled nerve cells in living, behaving experimental animals. These techniques can do much in the future to help us understand the functions of the brain and should serve well to address many of the new questions that will arise throughout the following chapters from mostly old, but as yet, partly undigested observations.

I will be pointing to areas of ignorance that can stimulate new questions. Firestein (2012) makes the point clearly in his book *Ignorance*: it is the bits we don't know that keep us awake at night and working in the lab. We need to be clear about what it is that we don't know. Sherrington, the leading neurophysiologist of his time, wrote: 'We must learn how to teach the best attitude to what is not yet known'.[4]

1.3 Seeing the brain as a tool for interacting with the world

I will treat the brain as a tool that we use in our interactions with the world. It is not the only one of our organs we use as a tool: our eyes, ears, noses, tongues, and palates also play an important role, as do our fingers, and even our legs, when we use them to explore the world. However, the brain is the key tool and without the brain the other parts of our bodies can do very little. They all depend on the brain. The brain is responsible for the ways we receive messages from the world and also contributes much to our reactions to these messages. We need to find out about the messages themselves and about how they are transmitted from one part of the brain to another. In the long term, we need to learn what contribution any one part of the brain makes in relating our actions to our perceptions, either in terms of an immediate reaction or over intervals varying from minutes to many years. We need to understand the role of our nervous systems in establishing our relationships to the world.

Once we recognize the brain as determining the ways in which we interact with the world, it becomes necessary to ask about the neural mechanisms that can produce any particular 'unconscious conclusion'. We need yet to learn how this happens in the young at first exposure as they learn to arrive at these conclusions, or when as adults they are first presented with entirely new sensorimotor relationships: perhaps when they first learn to reverse into a parking space, manipulate instruments in a mirror, or learn significantly new sensorimotor relationships after the regrowth of injured **sensory pathways** (e.g. Kaas et al. 1983).

---

[4] Cited by Eccles and Gibson (1979, page 24). Thanks to Lizzie Burns for pointing this out to me.

1.4 **The brain as a hierarchy of neural centres controlling our interactions with the world**

I will treat the brain as a hierarchy of brain parts, with each part monitoring lower levels, able to contribute to the control of their motor actions, as well as passing information to yet higher levels so that these, in turn, can monitor and control the ongoing activity at lower levels. This is an essential feature of all but the simplest brains. The cerebral cortex (see Figure 1.1(a–e)), which in its six-layered form (Figure 1.1(e)) is characteristic of mammals, forms a surprisingly large part of the human brain (Figure 1.1(a–d)). It covers the cerebral hemispheres, has many complex infoldings (Figure 1.1(d)), and represents several of the highest levels of the brain's hierarchies. The mammalian brain itself, including the cortex, can only act on the body or the world through the phylogenetically most ancient parts of the vertebrate **central nervous system**: the **brainstem** and the **spinal cord** (Figure 1.1(a–c)). These provide a two-way link between the brain and the body, and through the body to the world. Messages pass in one direction, literally down from the top (labelled 1 in Figure 1.1(a)) through the brainstem (labelled 5 and 6 in Figure 1.1(a) and 12–14 in Figure 1.1(b)) to the spinal cord (labelled 9 in Figure 1.1(c)). Messages leave the brainstem and spinal cord to pass along the peripheral nerves to muscles and glands for action. They pass in the opposite direction, up the hierarchy, to the more phylogenetically recent parts of the brain, formed by the cerebral cortex, for perception and cognitive processing: that is, for our conscious lives.[5]

---

[5] I will treat consciousness throughout this book as an abstraction that refers to a complex group of activities: perceiving, thinking, believing, imagining, remembering, and willing, and these are all verbs. Michael Billig (2013) has written an important book with the wonderful title *Learn to Write Badly*. In this he critiques the tendency for social scientists to use nouns in preference for verbs. For us, here, the point is that this tendency is not limited to social scientists. It is far more widespread than that and provides an irresistible opportunity for writing pompously about novel but non-existent entities when what is needed for clarity is simple writing about activities that are readily understood even by non-scientists. There is no identifiable thing that can be described as consciousness, no mysterious entity that lies hidden behind conscious acts and can be described by a noun.

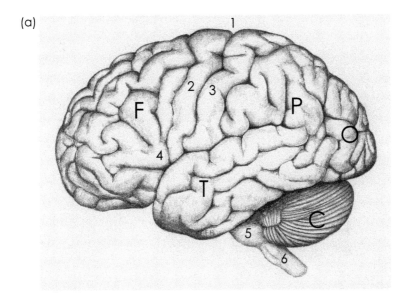

Figure 1.1 This figure has five parts to illustrate the great size and extent of the cerebral cortex, wrapping around much of the cerebral hemispheres and forming extensive infoldings into the hemisphere. The several parts of this figure also provide an overview of the cerebral hemispheres and their major cell groups, including the thalamus and the basal ganglia. The figures are planned to include primarily features that are relevant for understanding this book.

(a) Lateral view of a human brain, showing four lobes of the cerebral hemisphere, the frontal lobe (F), the parietal lobe (P), the occipital lobe (O), and the temporal lobe (T), each covered by a richly folded cerebral cortex. It also shows the cerebellum (C).

The frontal lobe is separated from the parietal lobe by the central sulcus (1), which also separates the precentral motor cortex (2) from the postcentral sensory cortex (3). (4) indicates the area described by Broca as involved in the production of speech (see Chapter 3). Two parts of the brainstem are shown, the pons (5) and the medulla (6). The midbrain is hidden by the large temporal lobe.

Illustration by Dr Lizzie Burns.

(b)

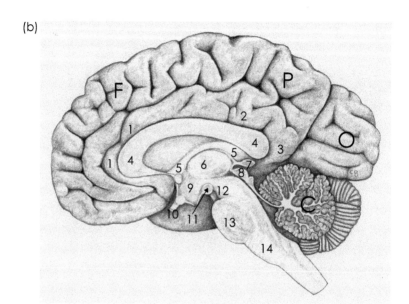

Figure 1.1 Continued

(b) A view of the medial surface of a human cerebral hemisphere. Note the small size of the thalamus (6) relative to the great expanse of the cerebral cortex (F, P, and O).

C, cerebellum; F, frontal lobe; O, occipital lobe; P, parietal lobe; 1, anterior cingulate cortex; 2, posterior cingulate cortex; 3, retrosplenial cortex; 4, corpus callosum; 5, fornix; 6, thalamus; 7, pineal gland; 8, superior colliculus; 9, hypothalamus; 10, optic nerve; 11,mamillary body; 12, midbrain; 13, pons; 14, medulla.

Illustration by Dr Lizzie Burns.

(c)

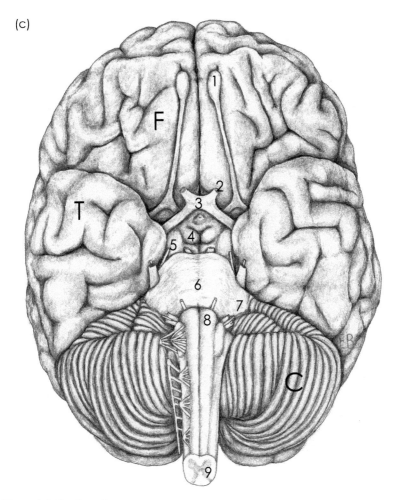

Figure 1.1 Continued

(c) A ventral view of the human brain showing the areas that rest on the base of the skull.

C, cerebellum; F, frontal lobe; T, temporal lobe; 1, olfactory bulb; 2, optic nerve (cut); (1 and 2 represent the first two of the cranial nerves, the rest emerge from the brainstem and have not been separately labelled); 3, optic chiasm; 4, mamillary body; 5, cerebral peduncle; 6, pons; 7, middle cerebellar peduncle (a bundle of nerve fibres entering the cerebellum from the pons); 8, the medullary pyramid; 9, spinal cord.

Illustration by Dr Lizzie Burns.

(d)

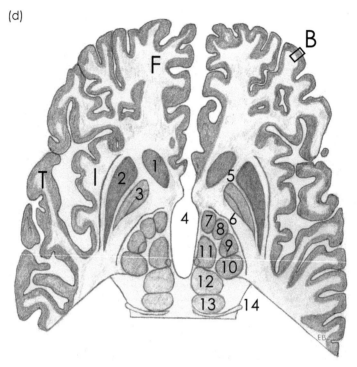

Figure 1.1 Continued

(d) A horizontal section through a human brain to show the richly folded cerebral cortex and some of the internal structures of the brain.

B shows a small area of cortex to illustrate the areas shown in Figure 1.1(e); F, frontal lobe; I, insula, a deep part covered by cortex never seen from the surface; T, temporal lobe.

1, caudate nucleus; 2, putamen; 3, globus pallidus (1–3 together form a major part of the basal ganglia; 4, third ventricle; 5, anterior limb of the internal capsule; 6, posterior limb of the internal capsule (5 and 6 carry the main fibre bundles to and from the cortex); 7, ventral anterior thalamic nucleus; 8, ventrolateral thalamic nucleus; 9, ventral posterior thalamic nucleus; 10, pulvinar; 11, dorsal medial nucleus; 12, superior colliculus; 13, inferior colliculus; 14, fourth cranial nerve, which supplies one of the six muscles that move the eyes.

The thalamus is subdivided into several separate functionally distinct cell groups, the thalamic nuclei: some of these play a significant role in this book and some of these are identified above. Others are shown in Figure 8.4. The present figure is based on horizontal sections at https:www.msu.edu/user/brains/human/horizontal.

Illustration by Dr Lizzie Burns.

(e)

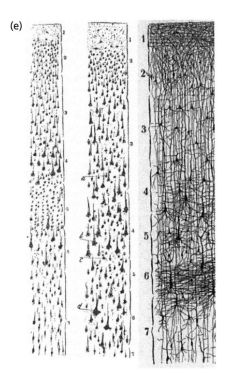

Figure 1.1 Continued

(e) Three sections through a small column of cerebral cortex (from Ramón y Cajal 1955) illustrate that different cortical areas have different structures, but that they all fit into a general plan based on a conventional six-layered pattern. This is the main basis on which the cortex has in the past been divided into different areas having different functions and different connections. The two sections on the left have been stained by a method, the Nissl method, which shows all of the nerve cell bodies in the thin section (probably only about 10–20µm in thickness) and their layered distribution in the cortex. Here, Ramón y Cajal, the leading neuroanatomist of his and probably of all time, has identified seven layers, whereas the contemporary view identifies six layers. (Ramón y Cajal's layer 5, characterized by small granular cells in the figure on the left, corresponds to the contemporary layer 4.) The left section was taken from the sensory cortex (labelled 3 in Figure 1.1(a)), receiving sensory inputs from the skin and deep tissues, which lies just behind the central sulcus. It has a strong granular layer 4 (Ramón y Cajal's 5). The middle section was taken from the motor cortex, which lies immediately in front of the central sulcus and gives rise to one of the most important motor pathways to the body and the limbs, and lacks an obvious granular layer 4. Note the markedly different structures of these two cortical areas.

The section on the right was from a post-mortem preparation from a 1-month old human infant and shows the cortex stained by another method, the Golgi method. This method stains relatively few of the cells and fibres in the tissue (generally about 1%), therefore the sections can be cut much thicker than for the Nissl stain. It stains not only the nerve cell bodies, but all of their processes, often completely. Each nerve cell has several processes that branch near the cell body, the dendrites, and another single process that commonly passes to another part of the brain before it forms extensive terminal branches that connect to other cells (shown in Figure 2.4 in Chapter 2, and Figure 5.2 in Chapter 5). Because so few cells are stained, each stained cell and its processes, show up against a relatively clear background.

There are no known reasons for thinking that any subcortical[6] part of any mammalian brain has the capacity to generate conscious processes. The cortical part of the hierarchy not only provides us with our conscious lives, but, as Chapter 4 demonstrates, all parts of the cortical hierarchy link through many pathways to the rest of the brain so that each can monitor and contribute to the control of lower centres. However, the lower centres also have significant capacities for complex actions on their own.

The lower levels of the brainstem and spinal cord provide our oldest and most basic sensorimotor interactions with the world. In our vertebrate ancestors, who lacked the characteristic six-layered cerebral cortex (see Figure 1.1(e)), these lower levels played a significant role in the control of many complex behaviours: for example, fish and frogs can catch flies, even through the refractive properties of an interface between water and air.

We need to learn how the cortex links to these complex older centres and how, in terms of known neural connections, each part of the cortex can contribute to particular aspects of our behavioural repertoires.

An example of a basic hierarchy can be provided by the way quadrupedal mammals like cats use their legs when they explore the world. A direct sensory-to-motor link is provided by the simple reflexes that control the muscles of the legs on the basis of incoming messages from the **sensory receptors** in the legs (Figure 1.2). On their own, limited to a single segment of the spinal cord, including only the input and the output, these reflexes cannot provide coordinated movements. However, if the whole of the spinal cord is intact, even if it has been completely separated from the rest of the brain, then, provided the body is adequately supported, the many rich connections within the spinal cord (not illustrated in Figure 1.2) can serve to let a cat walk on a treadmill, adjusting its speed as the speed of the treadmill changes (Grillner

---

[6] Subcortical refers to all parts of the brain that are not a part of the cerebral cortex. They all lie below the cortex in position and in the functional hierarchy. The recent suggestion by Feinberg and Mallatt (2016) that the midbrain may have provided some of our early vertebrate ancestors with conscious perceptions of the world is interesting but not easily verified.

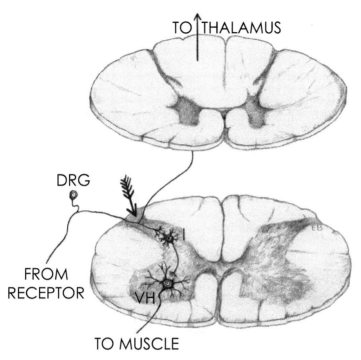

Figure 1.2 Schematic view of two sections through a human spinal cord, each showing the grey matter containing the nerve cells surrounded by white matter made up of large numbers of ascending and descending fibres connecting the different levels of the cord to each other and to the brain (only a single ascending fibre heading towards the thalamus is shown).

Each fibre is the outgrowth of a single nerve cell. A single incoming nerve fibre from a sensory receptor and an outgoing nerve fibre to the muscles are shown. The former comes from a dorsal root ganglion (DRG) cell that gives rise to two branches at the arrow. One branch ascends towards the thalamus and the other passes to a small interneuron, I, in the dorsal horn of the spinal cord, which, in turn, sends its axon to a motor cell in the ventral horn, VH, which innervates a muscle. When the DRG cell dies, both branches degenerate, the ascending branch as well as the local spinal branch.

Illustration by Dr Lizzie Burns.

2003). The basic sensorimotor controls of the lower locomotor mechanisms are intact at this spinal level.

Higher up, in the **midbrain** (labelled 12 in Figure 1.1(b)), is a centre (not illustrated) that has been called the mesencephalic (i.e. midbrain) **locomotor centre** (Lee et al. 2014; Sherman et al. 2015). This can influence the initiation or cessation of locomotion at spinal levels and, in turn, receives inputs from other regions of the brain including cortical and subcortical motor centres of the cerebral hemispheres (the **basal ganglia**, labelled 1–3 in Figure 1.1(d)) as well as the **cerebellum** (Figure 1.1(a–c)). The details of the connections are not yet well defined although it is clear that the cerebral cortex, the basal ganglia, and the cerebellum, including several levels within the cerebral cortex itself, play an important role in the control of locomotion.

Although we know that many such complex hierarchies exist, involving all levels of the brain, with each level able to record the events at a lower level as well as having the outputs that can contribute to the actions at the lower levels, the detailed structures of the hierarchies still need to be defined. Some of them are described in more detail later in this book, but none can be thought of as clearly understood as a whole or in terms of their interactions with other levels. One purpose of later chapters is to draw attention to many of the details about which we need much more specific information.

I will stress that there is still much that is currently unknown and unexplored about the structures and functions of many parts of the brain. This can seem depressing, but it should be recognized as an important challenge, a challenge that forms a significant part of the following chapters. For many areas of the brain we still need to learn not only what it is that one particular group of cells contributes to our lives, but also to find out how these nerve cells interact with cells in other parts of the brain to produce a specific result.

We need to identify the questions that we should be asking about each brain part. These questions seem clear for the parts of the brain that are close to the sensory inputs or the motor outputs. Thus, for any one motor output we need answers to questions about how neural activity in the brain gets distributed to muscle groups needed for specific actions. For this we have some clear answers. Similarly, for any one sensory input, visual, auditory, gustatory, and so on, we know that we have to start by varying the sensory stimuli so that we can learn how specific sensory events are represented by neural activity at any one level of the sensory pathways.

However, as we move centrally, many parts of the brain provide no clear clues as to what questions we should be asking, or else provide opportunities for many diverse questions that are liable to produce an excess of possible answers. I use Chapter 6 to demonstrate this for a group of pathways that lead through a thalamic **relay** to three related areas of the cortex. For these pathways, we have some clear answers for one small component, and no clear information at all for the two largest components. The answers we get depend on the ways in which we frame our questions, and the further we are from the sensory inputs or the motor outputs, the more difficult it is to find the appropriate questions.

1.5 **The cerebral cortex and the thalamus**

The **thalamus** and the six-layered cerebral cortex (Figure 1.1(d),[7] labelled 7–11, and Figure 1.1(e)) form an important part of the hierarchies of the previous section. This combination of many highly differentiated and functionally distinct cortical areas that together with their connections with the thalamus form a reciprocally connected hierarchical thalamocortical system, characterizes the mammalian cortex and distinguishes it from the brains of our non-mammalian vertebrate ancestors. Not only is this system large and dominant in many mammals, but it appears to have enlarged independently in many different mammalian lines. Most particularly, it dominates the rest of the brain to a significant extent in primates and in us; it is possibly the part of the brain that we are furthest from understanding and it is the part on which our conscious lives depend.

The cortex receives essentially all of its inputs about events in the brain, the body, or the world from the thalamus, a structure containing many nerve cells arranged as several functionally distinct groups. Some of the **thalamic nuclei** are shown in Figure 1.1(d) (labelled 7–11). These together form a structure about the size of a walnut in the human brain, or of a small pea in a rat. The thalamus is surprisingly small in relation to the vast extent of the cortex. It has been described as a gateway to the

---

[7] This illustration only shows a few of the many thalamic nuclei.

cerebral cortex, and although it serves as a way into the cortex, there is a different way out from the cortex.

The cortex is reasonably regarded as highest in the hierarchies and is itself organized in monkeys as a hierarchy of five, six, or more levels (Van Essen et al. 1990, 1992), probably more in humans, of separate cortical areas, with higher areas monitoring the outputs of lower areas, contributing to their control and also able to contribute more directly to lower centres that are directly involved in the control of ongoing movements (Figure 1.3). Only the earliest levels of the cortical hierarchy are shown in Figure 1.3.

One key feature of the cortical hierarchy is that some inputs to the thalamus for relay to the cortex come from lower centres other than the cortex; these supply the '**first-order' thalamic nuclei** (labelled 'FO' in Figure 1.3). Other inputs come from the cortex itself. These supply '**higher-order' thalamic nuclei** (labelled 'HO' in Figure 1.3) and serve as the first stage of transthalamic corticocortical connections. These higher-order transthalamic connections are beginning to be explored but at present they still represent many challenges regarding the connections that they form and the messages that they carry. A second key feature is that essentially all of the **axons** that bring messages to the thalamus for relay to the cortex are branched (see Chapter 9 and 10 for more detail). One branch goes to the thalamus for relay to a higher area, and the other goes to centres that serve to control actions: primarily the muscles. It is these branched connections that allow higher centres not only to monitor what a lower centre is doing, but also to contribute, when necessary, to the relevant motor response.

The thalamus is widely recognized as a 'relay' to the cortex. This is useful, but it should not be seen as though the thalamus serves as a simple switch or booster point in an electrical circuit just passing on signals. It is often important to remember the early meaning of a 'relay', when it referred to a relay inn, where horses were changed, travellers stretched their legs, took some refreshment, perhaps spent the night, met other travellers, and often got involved in interactions that could sustain a large part of a Victorian novel. We will see that the thalamic relay is complex. That complexity is at the core of this book.

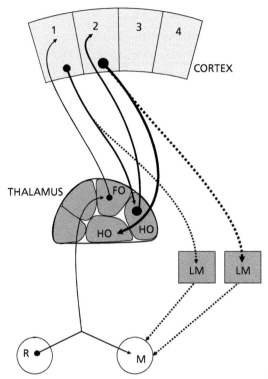

Figure 1.3 Schematic representation of the branching patterns of axons that bring messages to the thalamus for relay to the cortex. FO, a first-order thalamic nucleus is receiving sensory information from the sensory receptor (R), which also innervates the cells that innervate muscles (M); HO, two higher-order thalamic nuclei are receiving input from axons that arise in the cortex and sends a branch to a lower motor centre (LM). Notice that the three inputs to the thalamus, that from the receptor and those from cortical areas 1 and 2, are shown at increasing thicknesses to represent the increasing information carried to the higher cortical monitors. Only the first three levels of the cortical hierarchy are shown.

If we want to understand the information available to the cortex, we have to understand the messages that reach the thalamus and how they are relayed to the cortex. We have to understand which messages are passed on to the cortex at any one time, and which messages are not passed on. That is, we have to understand the **thalamic gate**.

At any one time, we are only aware of a very small proportion of the messages that reach our thalamus for relay to the cortex. This is due to the action of the thalamic gate (and is discussed further in Chapter 8).

In order to understand the mechanisms that prevent the relay of many messages, we have to understand the functional interconnections between thalamus and cortex, to learn about the actions that any one cortical area can have, directly or indirectly, on the thalamic gate, and to learn about other pathways that can silence the thalamic relay. These issues are considered in Chapter 8, and have been examined in more detail by Sherman and Guillery (2013).

We do not know where in the hierarchy that supervises lower centres, particularly where in the cortical parts, the unconscious conclusions that lead to the constancies of perceptions about the world are generated. This hierarchy also plays a role in controlling our actions, memories, thoughts, and feelings, and is discussed in Chapters 12 and 13. It dominates the rest of the brain. We may well be on the threshold of understanding its functions and we will find that further insights will depend on learning a great deal more about the details of the connections that allow the higher cortical areas to supervise and control the lower areas.

1.6 How simple organisms interact with the world

In order to explore the nature of the interactions between our brains and the world at any hierarchical level, it is useful to start by looking at the most basic level in an extremely simple organism and asking how such an organism, something more or less like one of the earliest of our single-cell ancestors, related to the world.

The sensorimotor capacities of bacteria have been studied in great detail (Adler 1966; Berg 1975) and we can treat these as our model. Bacteria have sensory receptors that are part of a mobile **cilium**, a long, thin, tiny organ whose movements can move a small organism through its fluid environment. Stimulation of these sensory receptors can produce movements of the cilium, and through the cilium, of the organism. The action of these sensory receptors can produce a pattern of movements that can either move the organism up the gradient of concentration of a stimulant (perhaps glucose or light) or down the gradient (perhaps a toxic acid or light in an organism that thrives in the dark).

Our early vertebrate ancestors were too large to be moved by a mere cilium. Their receptors remained on a cilium or on a structure developmentally derived from a cilium, and many of our receptors—visual, auditory, olfactory, and gustatory—still form structures closely related to an early cilium, although many are now non-motile. The receptors of vertebrates need to be linked by nerves to the spinal cord (Figure 1.2), the brainstem, or to the brain itself for the pathways that deal with vision and olfaction. Eventually the messages must pass to muscles (or glands) in order to produce the actions of larger animals.

Figure 1.2 shows at the large arrow that the close sensorimotor link of the cilium remains: all of the signals that come from the periphery have one primary purpose—to produce, or contribute to, a specific movement. All of the sensory inputs pass along the local branch in the nervous system, eventually for an output to a muscle (or gland) for action. That is what the sensory nerves are for. They serve to initiate our interactions with the world. The messages that pass along the ascending branch to higher centres, including the thalamus and cortex, are all conveying a message about such interactions. Messages that are passed to daughter branches at any axonal branch point are both essentially copies of the message passing along the parent branch (Cox et al. 2000; Raastad and Shepherd 2003). Whatever 'view' of their world these earliest vertebrates received from the ascending branches was a 'sensorimotor' view. It was a view based on their interactions with the world: this ascending sensory message is about events at the sensory receptor and is also about instructions for specific movements already on their way to the muscles. One message with two meanings travels up the ascending branches. However, the identification of the stimulus, as glucose or light or anything else, was not needed at the earliest stages; only the appropriate interactions were relevant for survival, and these interactions are still the raw material of our conscious lives. All of our sensory inputs serve to produce motor actions.[8]

---

[8] They can be blocked in the thalamus, and the reactions can be inhibited in many different ways.

When, during later vertebrate evolution, the spinal sensorimotor levels started to send messages up to higher levels of the brain, the messages did not come as pure messages from the sensory receptors dissociated from the motor actions: they came along branches as a part of the spinal sensorimotor link reporting to higher centres about the sensorimotor interactions. That is, they carried messages to the thalamus about our interactions with the world, directly linking current information about the world to the immediately related instructions for necessary movements. Depending on the neural connections and on the physical properties of the parts that were to be moved, the message reaching a higher centre might anticipate or follow the actual movement, but the ascending message always included information about both the output of the sensory receptor and the instruction for movement. Of course, both reached the higher centre at the same time (it was a single message with two meanings) before any message about the sensory *outcome* of the action (called the **reafference**) could arrive. The initial single message was about two events: one sensory, about environmental, bodily, or neural events in the immediate past; the other, about the instruction for a motor action in the immediate future. This single message would be followed by a message about the sensory outcome actually produced by the action (the reafference). That is, the higher centres received information that anticipated actions and their sensory consequences.

Many, possibly all (discussed in Chapters 9 and 10), of the messages that reach the thalamus for relay to the cortex come along branched nerve fibres, comparable to the ascending branch shown in Figure 1.2. They send one branch to the thalamus for relay to the cortex and another branch to a group of nerve cells in one or another neural centre concerned with the production or control of motor actions.

The functional implications of these branching patterns seem clear: our knowledge about the world is based on our interactions with the world. We experience the world by interacting with it, not by undertaking objective observations of it as though in our daily lives we were scientists concerned with defining the pure sensory components that reach us from the world as light, sound, or heat and so forth so that we can decide how to act. The idea that first we perceive the world, and then the cerebral cortex acts on the basis of these perceptual inputs, computes the needed outputs, and subsequently sends instructions for movements to the muscles, is an old and well-established view. It fits well with our intuitive sense of how we relate to the world and it has long been taught in neuroscience courses. I have called it the standard view of perception, where messages relate sequentially to distinct events and are interpretable as relating to single items: sensory events, motor instructions, intermediate computations, or items for memory storage. In the standard view, messages about motor instructions follow the sensory inputs and pass along different nerve fibres. I will suggest that this does not accord with the neural facts. This view needs to be replaced by a view of single messages that can be about sensory events *and* motor instructions: a sensorimotor view of messages that may have more than one meaning in terms of the events or actions represented. This is a complex, or sensorimotor, view that is fundamentally different from the standard view.

1.7 A clinical illustration of sensorimotor links in perception

MH Romberg, who in the late 1840s and early 1850s published one of the earliest textbooks of clinical neurology (Finger 1994), described a clinical condition, **tabes dorsalis**, that destroys the cells of the spinal **dorsal root ganglia** (see Figure 1.2). Because the nerve cell bodies are affected, the peripheral processes carrying inputs from sense receptors as well as the central processes of the dorsal root ganglion cells die. This leads to marked sensory losses including those that record the movements of the limbs. Patients lose their ability to identify or distinguish sensory stimuli applied to the affected parts of the skin or deep tissues, and are also unable to judge the position of their limbs if their eyes are shut. The simple action of closing one's eyes and touching one's nose with one's finger becomes difficult if the arms are affected because the joint and muscle receptors that signal limb movements are no longer functioning.

Romberg also described a test that is used not only by clinicians for patients but also by police forces for testing drunks. For **Romberg's test**, the subject stands with the feet close together and then is asked to close both eyes. A subject with losses of the lower dorsal root ganglia, or a drunk, will tend to lose their balance and the examiner has to be ready to prevent a fall. Romberg described this loss as a motor loss, a point made by Spillane (1981), who wrote, '... it was not, even then, necessary to push the classification he adopted to such an absurd degree ... to consider tabes dorsalis as a motor disability, despite his description of its essential element— a sensory ataxia'. This is an interesting and important comment. Spillane treated the deficiency demonstrated by the test as a *sensory* loss that produced abnormal movements (a sensory ataxia), whereas Romberg treated it as a *motor* deficit. On the basis of the connections, we should regard it as a sensorimotor deficit.

The issue was addressed later by Brodal (1981), who pointed out that the degenerative changes involve not only the fibres that ascend for relay to the thalamus and cortex, which was a key consideration for Spillane. The degeneration also includes the fibres that enter the spinal cord and innervate the local reflexes involving a final motor output through the **ventral root** to the muscles (Figure 1.2). Brodal drew attention to the fact that the fibre losses in the ascending, sensory fibre bundles are readily seen in post-mortem material, because all of the degenerating fibres are closely bunched together. In contrast, the degenerating motor fibres that (directly or indirectly) innervate the **ventral horn** lie among many other, unaffected normal fibres so that their loss or degeneration can only be seen with difficulty or not at all. That is, the post-mortem anatomy strongly suggests a loss in the tabes patients of the sensory pathways on the way to the cortex for perception but shows little or no evidence of the fibre loss that produces the motor changes seen after the dorsal root ganglion cells are lost, and this fits with the marked sensory losses.

Brodal also describes clinical as well as experimental evidence from monkeys showing that damage to the ascending fibres in the spinal cord on their way towards the thalamus *alone*, leaving the motor branches intact on their way to the ventral horn cells, does not show the full ataxia of the tabes lesion although the sensory losses are, of course, the same. Romberg's sign is absent after such a pure sensory lesion: this sign is not a 'sensory ataxia'. The classical tabes lesion is thus to be considered as both sensory and motor. This is a crucial point for understanding the functions of the spinal branches that arise from the incoming dorsal root axons as they pass on their way towards the thalamus (Figure 1.2 at lower arrow).

We will see in Chapter 2 that Bell and Magendie (see Finger 1994) had earlier identified the **dorsal roots** as sensory and that the ascending fibres had been traced through the relevant relay stations to the thalamus for relay to the **sensory cortex**. This set the stage for the view of a sensory ataxia, because there was a long tradition that the way into the brain is separate from the way out. Although the local spinal branches have a motor action that continues to contribute to the stability of the patient's stance when only the ascending axons are damaged, the traditional view saw the motor outputs as something separate from the functions of the dorsal root ganglia, not relevant for a positive Romberg's sign, which was erroneously described as a 'sensory ataxia'.

The messages that the thalamus receives from the ascending fibres and relays to the cortex can be seen as providing, in a single axon, what O'Regan and Noë (2001), in an important summary review of perception, have called 'sensorimotor contingencies'. These authors presented extensive evidence that perception often depends on actions, and that action and perception may often be inseparable.

For many of the major sensory pathways, we currently lack information about the actual functions of the relevant motor branches. The questions have generally not been asked. They appear to be irrelevant in the standard view, but can be recognized as a crucial piece of information on the interactive view.

Chapter 2

The pathways for perception

2.1 Summary

Chapter 2 outlines some of the evidence on which the seemingly strong standard view has been based. The early discovery that ventral nerve roots of the spinal cord provide a motor output and dorsal nerve roots provide a sensory input supported the dichotomy of the standard view. Then as each sensory pathway was traced to the thalamus for relay to the cortex, the separate inputs from the sensory receptors—visual, auditory, gustatory, and so on—could be seen as providing the cortex with a 'view' of the world. The nature of this view became strikingly clear once investigators could understand (read) the messages that pass along the nerve fibres on the basis of very brief changes in **membrane potentials**, the **action potentials**. However, many branches given off by sensory fibres on their way to the thalamus remain unexplained in the standard view. These are important for the integrative sensorimotor view and their precise functional roles need to be defined.

2.2 The 'standard' view linking the brain to the world

The standard view of how our perceptions relate to our actions is based on one of two related views that are often treated separately. Both were clearly expressed by Descartes (1662) and are both present in Figure 2.1.[9] One, the mind–body or dualist view, represents

---

[9] Wheeler (2007) considers the independent nature of these two views that can both be termed Cartesian. The point is important because many neuroscientists who do not regard themselves as dualists still hold firmly to this other important Cartesian idea of how actions arise from perceptions.

our mental functions, localized as the soul or the mind in the **pineal gland** (labelled P in Figure 2.1), as separate from the body and the world. The other view present in this figure is the simple, standard view, which links perceptions to actions through the pineal. Today, the standard view is commonly treated as quite independent of views about the mind.

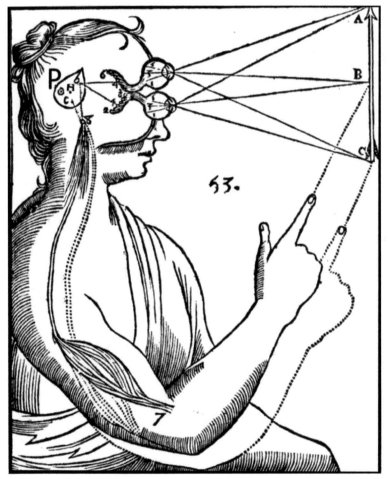

Figure 2.1 This figure shows a schema from Descartes (cited by Finger 1994) relating the incoming sensory messages coming from a physical object, represented by the arrow labelled A, B, C, and going to the pineal gland (P) where a mental representation is formed. Messages are then sent from the pineal gland along the nerves to the muscles for appropriate motor outputs.

Figure 2.1 shows messages, here coming from the eyes, going to the pineal gland (P)[10] for perception, to then be sent along the nerve fibres to the muscles for action. The mind, in the pineal, in the figure links the inputs to the outputs. This standard view does not readily fit the hierarchy of centres monitoring our sensorimotor interactions with the world, which was introduced in Chapter 1. It is the sensorimotor view that lies at the heart of the rest of the book.

Figure 2.2 illustrates a contemporary version of the standard sensory-to-motor view. A part of the brain often replaces the pineal (e.g. Crick and Koch 2005), and cognitive processing, which is thought to occur in the cerebral cortex (among the arrows at the top of Figure 2.2, not in the pineal), may replace the mind, but that does not alter the conceptual distinction between the standard view about action and perception, on the one hand, and the dualistic mind–body view, on the other. They are generally treated as separate issues.

We have to look carefully at both of the proposed relationships: those of the physical to the mental and those of the sensory to the motor. In this chapter we look at the sensory-to-motor dichotomy of the standard view. This view has previously been challenged by a number of authors (Merleau-Ponty 1958; Gibson 1986; Varela et al. 1993; Churchland et al. 1994; Clark 1998; Hurley 1998, 2001; O'Regan and Noë 2001; Thompson and Varela 2001; Pfeiffer and Bongard 2007; discussed in Chapter 12) who see evidence for close interactions between perceptions and actions, sometimes described as an embodied view of perception or an interactive view. However, generally the challenge has been based on perceptual or behavioural evidence, not on specific neural connections, and these are relevant here. We will look at both views, the standard and the interactive, in terms of known neural connections and must look at these connections in some detail because they can support both views. However, only the interactive view allows us to see the functions of the hierarchy of connections that monitors our sensorimotor interactions, introduced in Chapter 1 and crucial for Part IV.

---

[10] Descartes nominated the pineal gland as the seat of the soul or mind partly on the basis that it needed to be a single midline structure.

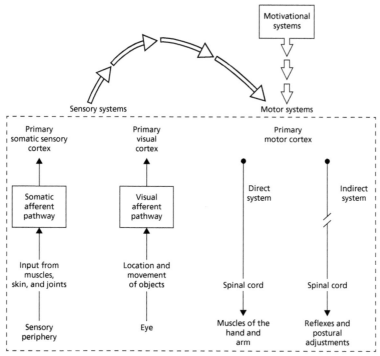

Figure 2.2 This figure is a schematic view based on a figure in Kandel et al. (1991) showing how 'most behavioural acts' are often thought to relate to the cerebral cortex.

Reproduced from Eric R. Kandel, James H. Schwartz and Thomas M. Jessell, *Principles of Neural Science*, 3e, Figure 19.6, page 281, McGraw-Hill Education © 1991, McGraw-Hill.

Much of what we currently teach students about the brain has its *conceptual* origins not in the neural facts themselves, which in practice determine the way in which we perceive the world, but in an intuitive and well-defined view established since biblical times about a real world that is out there and was created for us to see and experience. This leads to studies about messages from the, supposedly, known world that are passed along the neural pathways to the cortex. These studies use the laws defined by physicists and chemists about light waves, sound waves, odours, and other physical forces to show how such a world can act on the brain to produce our perceptions of the world.

Neuroscientists have often in the past interpreted known neural functions and connections of the brain starting from such a standard view about the world, seeking on this basis to understand the workings of the brain and its parts. However, since our view of what the world is like depends on the ways in which we receive information about the world through our nervous systems, it is worth turning our approach to the problem around and studying first how our nervous systems link us to the world. Then we can raise questions about how this relates to what we can expect to know about the world. This follows well behind and in the footsteps of many others—including, as indicated in Chapter 1, of Helmholtz and Hume—but instead of basing the analysis on understanding perceptions and then asking how that determines what we can know about the world, I will be asking about the extent to which our knowledge of the nervous system can help us to understand what we can expect to know about the world once we begin to understand the messages that pass along the nerves.

The view of the brain prevalent when I attended an introductory course on neuroanatomy as a second-year medical student in 1949, was of ascending pathways carrying messages towards the cortex and descending pathways, as shown in Figure 2.2, carrying motor instructions away from the cortex to the brainstem and spinal cord for transmission to muscles and glands. The focus was on the major sensory and motor pathways. The functions of many other brain parts and pathways were unknown or poorly defined and even today the functions of many parts of the brain are, as we will see, still unknown. Even defining what we mean by the functions is not always clear: do we mean *what is lost in the capacities of the organism when the pathway is lost, abnormal, or damaged?* Do we mean *what happens in terms of behaviour when the pathway is actively transmitting messages from one cell group to another?* Or do we mean *that we understand the nature of the messages that are being transmitted and how those messages are processed?* These are very different questions about the functions of a pathway in the brain and there are relatively few pathways for which we have answers for all of them (for examples, see Chapter 6).

These three distinct questions about the functions of pathways, and many other questions, were generally not clearly defined for us students as we struggled to understand the brain and what it can do; the main focus was on the first question about injuries and abnormalities, which, of course, is most relevant for a medical student. I have italicized these questions here so that they can serve as an introduction to some of the major problems and concepts that today still need to be recognized, discussed, and explored for many pathways as our knowledge of the brain increases. They are particularly relevant for sensory pathways that bring information about the body and the world to the cortex.

2.3 **The way into the cortex**

An early step for breaking the code of the inputs to the brain was the recognition, by Bell in 1811 and by Müller in 1826, that each sensory nerve must be transmitting a specific and limited type of sensory message. Charles Bell wrote: '... while each organ of sense is provided with a capacity for receiving certain changes to be played upon it, as it were, yet each is utterly incapable of receiving the impression destined for another organ of sensation' (see Finger 1994, page 135). That is, we cannot sense changes in temperature with our eyes or ears and cannot see light with the receptors in our skin. Localized pressure applied to the eye produces a visual sensation in spite of the absence of a visual stimulus. Johannes Müller introduced the concept of 'specific nerve energies' to describe this specificity of the messages in any one axon and today it is often treated in terms of **labelled lines** formed by our neural pathways. The pathway along which the message arrives is itself an important part of the message but represents by no means all of it. If we had a separate telephone line for each of our colleagues we would not need to ask who is calling and we would have a fair chance of anticipating the message, but would often still need to listen to what was being said.

2.3.1 **Understanding the messages**

Efforts to understand the messages that are transmitted from the peripheral sensory receptors through the peripheral nerves and the long ascending spinal and brainstem pathways to the thalamus and cortex have a long history. Many early students of the brain thought about the messages that must pass along nerve fibres, but the idea that one could tap into a nerve fibre or a nerve cell and read the message had to wait until the 1920s, when the very brief changes in the electrical potential across the membrane surrounding each nerve cell or fibre could be accurately recorded. These 'action potentials' (Figure 2.3(a–c)) travel rapidly (up to 100m per second) along nerve fibres. They originate in the part of the nerve cell, the **axon hillock**, from which the axons leave the cell (Figure 2.4) and then pass along the axons, for varying distances: in large animals these vary in length from a metre or more to small fractions of a millimetre.

Action potentials represent a major way for nerve cells to communicate with each other, particularly when the communication is over significant distances and needs to be rapid. The messages are passed from one cell to another at specialized junctions called **synapses** (labelled b in Figure 2.4) (see Chapter 7 for a discussion on synapses).

In 1949, answers for the third question raised in the previous section, the nature of the message, were just becoming clear for some of the sensory pathways of the brain. Studies by Adrian in the 1920s (Figure 2.3(a)) and other later studies (e.g. Figure 2.3(b)) showed recordings of action potentials passing along peripheral nerves and carrying information about sensory events from the body to the brain, rather as a Morse code carries messages, but using dots without the dashes and relying on the temporal distribution of the dots to interpret the dynamics of the message (Adrian 1928, 1947). The major publications that defined how stimuli—to skin, hairs, joints, eyes (Figure 2.3(c)), or ears—related to particular patterns of action potentials in the central nervous system itself, particularly in cortical or thalamic cells or in pathways on the way to the thalamus, were not published until a few years after I had taken the introductory course in 1949. I remember the sense of excitement when we saw some of the

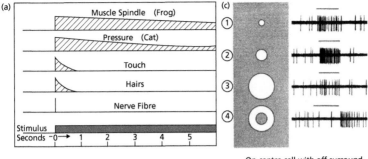

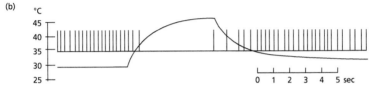

On-centre cell with off surround

Figure 2.3 This figure shows records of sensory messages in the peripheral and central nervous system.

(a) Recordings obtained from peripheral nerves in response to stimulation of different sensory organs and of a nerve fibre itself. The records show how the frequency of the recorded action potentials changes in relation to the stimulus, which is shown as continuous throughout the recording (stimulus). The individual action potentials are not shown; the height of the line represents their frequency.

Reproduced from Adrian, E.D., *The Basis of Sensation: The Action of the Sense Organs*, p. 79, Figure 19, WW Norton & Co., New York, 1928.

(b) Recordings obtained from a single nerve fibre from a nerve (saphenous nerve) in the leg of a cat, showing the individual action potentials recorded as the skin of the leg was warmed from 30°C to 45°C and then cooled again. The total time of the record shown was just over 20 seconds.

Reproduced from *Pflüger's Archiv für die gesamte Physiologie des Menschen und der Tiere*, Afferente Impulse aus der Extremitätenhaut der Katze bei thermischer und mechanischer Reizung, Ingrid Witt and Herbert Hensel, 268 (6), p. 586, Abbildung 3b, DOI: 10.1007/BF00362294 © Springer-Verlag Berlin Heidelberg 1959. With permission of Springer.

(c) Recordings obtained from the thalamic visual relay (the lateral geniculate nucleus) of a cat. The cat was viewing a screen and the circular visual stimuli shown on the screen are shown on the left. The stimulus appeared on the screen during the period shown by the line above each record, for 1 second in recording 1. Note how the temporal distribution of the action potentials depends on nature of the stimulus.

Reproduced from D. H. Hubel, T. N. Wiesel, Integrative action in the cat's lateral geniculate body, *Journal of Physiology*, 155 (2), p. 387, Figure 2, DOI: 10.1113/jphysiol.1961.sp006635 © 1961 The Physiological Society.

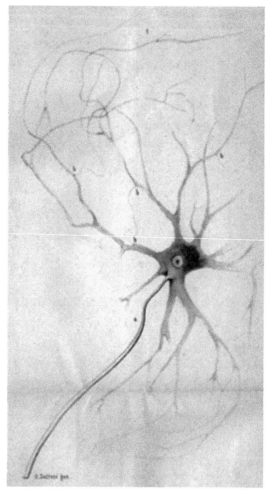

Figure 2.4 A large nerve cell (30–50μm in diameter) from the spinal cord (of a cow or sheep). Several sinuously branching dendrites are continuous with the cell body. Deiters (1865) showed that these dendrites are quite distinct in their structure (and in their reaction to manipulations, to chemicals and dyes) from the single, smooth refractive axon (a) that arises from a pale swelling of the cell called the axon hillock. Cells such as this, the ventral horn cells of the spinal cord, are today known to send their axons out of the spinal cord and form the final outputs for action. The cell was dissected manually under the microscope. Deiters suggested that for nerve cells generally, the dendrites might represent the part of the cell receiving inputs from incoming nerve fibres such as those shown by b, and that the single axon, characteristic of the many different cell types he had dissected, would represent the output of the cell.

From Deiters (1865).

early reports (e.g. those shown in Figure 2.5).[11] These observations formed no part of our course in 1949. They followed more than a century of clinical and anatomical studies that had defined the sensory pathways and it is important to look at some of that history in order to understand the strength of the standard view.

There is one crucial point about the messages that nerve cells transmit over long distances by means of action potentials travelling along the axons: events occurring in the body or the world often include changes that occur in milliseconds. We move our eyes three to four times per second and each action is rapid. We can follow a ball travelling rapidly and adjust our limbs to catch it accurately. In order to be useful for recording events in the body or the world we need messages to be transmitted from one nerve cell to another as discrete events and rapidly. This has led to an important distinction between messages that can serve to represent events in the body and the world and messages that are too slow to serve this purpose. For sensory pathways that involve the thalamus and cortex, Murray Sherman and I (Sherman and Guillery 1998) have suggested that there are two types of neuronal connection (discussed in Chapters 8 and 9): **drivers** can act rapidly and transmit information about events in the body and the world generally, whereas **modulators** take large fractions of a second or even longer, to pass messages from one nerve cell to another. They can change the way a message is transmitted (see Chapter 8) or they can transmit messages about slower events, such as changes in mood, changes in sleep–wake states, or changes in the controls of our internal organs but they cannot transmit information about the usual rapid changes in our relationships to the world.

---

[11] Once I moved into a teaching post, in 1953, I joined the members of the preclinical departments for tea in the medical library (the 'Thane Library'). The newly arrived journals would be available for us to look through as we consumed our bread and jam and tea! (I don't think that the jam got on the books to a significant extent.) Those who did not feel inclined to talk would look through newly arrived journals, and then some new article might stimulate lively conversation. I have a memory (possibly a false one stimulated by later events) of coming across these figures one teatime, and getting completely lost in details of the figures.

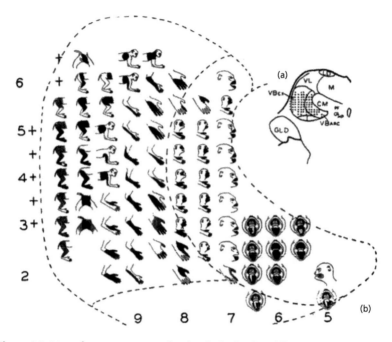

Figure 2.5 Map of responses to tactile stimuli obtained at different points in the ventral posterior thalamic nucleus (here called the ventrobasal nucleus (VB) in (a)) of a monkey. The stimuli were applied to the body parts of the monkey shown in (b) and the placement of the figurines shows the part of the nucleus where responses were obtained. Recording electrodes were lowered through the nucleus from the top of the nucleus to the bottom and the figure shows 12 such electrode penetrations.

Reproduced from Vernon B. Mountcastle and Elwood Henneman, The representation of tactile sensibility in the thalamus of the monkey, *Journal of Comparative Neurology*, 97 (3), p. 421, Figure 3, DOI: 10.1002/cne.900970302, Copyright © 1952 The Wistar Institute of Anatomy and Biology.

In his 1928 book, *The Basis of Sensation* (page 91), Adrian wrote: 'We will assume that the central nervous system is able to get every scrap of information out of the message, or let us say, everything that could be learnt by a physiologist who could isolate each nerve fibre and could record the impulses in it'.

This comparison between the capacities of the physiologist and those of the brain raises questions for both the physiologist and the brain. So far as the physiologist is concerned, when the recordings were obtained from anesthetized animals, as they generally were at the time and for many years later, the physiologist's knowledge was limited because the brain would be acquiring only a part of the information that the animal normally uses in its interactions with the world. Specifically, a physiologist in Adrian's time would not be looking for an interpretation of the message that might relate to an instruction for an upcoming action or to an anticipated change in the sensory input (the reafference).

Whereas the physiologists depend on their knowledge about the stimulus and its changes with time,[12] the brain lacks this information. The brain only has the temporal distribution of the action potentials along any one labelled line to extract information about the dynamics of the event that produced the action potentials. That is all that the brain receives. The brain receives nothing other than the action potentials and these provide the raw material for learning our 'unconscious conclusions' about the world. In order to know the world, we use the brain as a tool to interrogate the dynamics of the incoming action potentials in terms of the sensorimotor interactions that occur over time, we have no other way. The sensory message on its own can have little meaning for the brain. In our daily lives, each message must be related to the anticipated action and its sensory consequence (the reafference) so that the constancies of our relationships to the world can be learnt.

2.3.2 Tracing the messages towards the cortex

In the early nineteenth century, a law about the flow of information into and out of the spinal cord was formulated by Bell in 1811 and experimentally demonstrated by Magendie in 1822. This law, often called the **law of Bell and Magendie**, stated that sensory messages from the body enter the spinal cord through the dorsal roots and that motor messages pass from the cord to the muscles through the ventral roots (see Chapter 1, Figure 1.2).[13] This gave the relationship between inputs (for

---

[12] An important part of a physiologist's report has to include the details of the stimulus, it nature, and its timing. These are not directly accessible to the brain.

[13] It is relevant that all of the inputs to the brain except the visual and olfactory inputs enter through the spinal cord or brainstem and that the law of Bell and Magendie extends to the brainstem.

later perceptual processing) and outputs (for actions) a clear separation and distinct directional basis. A search then followed in the later parts of the nineteenth century for the further necessary connections linking the spinal dorsal roots to the thalamus and cortex. Also, in turn, the way out from the cortex to the spinal cord needed to be linked to the nerve cells of the ventral horn of the spinal cord and thus through the ventral roots to the muscles and glands. These studies can be seen as a search for the separate sensory and motor pathways proposed in the Cartesian scheme.

The ascending pathways of the spinal cord had an interesting history. Sometime before these had been traced to the thalamus, spinal injuries were known to interrupt sensory perception from the body and the limbs, but there was a problem about localizing the relevant pathways in the spinal cord. Brown-Séquard (1868, 1869), studying spinal injuries, first experimentally in frogs and later in patients that had suffered such injuries, concluded that some of the sensory pathways cross within the spinal cord on their way up to the brain whereas others do not cross in the cord. He reported that when one half of the spinal cord is damaged in a patient, there is a loss of tactile sense and of the sense of limb position below the level of the damage on the same side of the body. In addition to this, there is a loss of pain and thermal sense on the opposite side together with loss of some tactile sense. He concluded that two independent pathways with distinct functions ascend through the spinal cord. Both pathways were later traced to the thalamus on the side opposite to the lesion. They were shown to cross at different levels, were recognized as having distinct functions, and were given different names: medial lemniscal pathway (or posterior **column** pathway) for one, and **spinothalamic pathway** (or ventrolateral pathway) for the other. This was an early recognition that different pathways carry functionally distinguishable messages, extending the concept of specific nerve energies from the peripheral nerves to the pathways of the central nervous system.

It took many detailed anatomical studies during the second half of the nineteenth century to demonstrate these pathways, tracing their course all the way from the spinal cord to the cortex, with relays in the spinal cord, the posterior columns, and the thalamus. These were studies of post-mortem material from patients who had died after spinal injuries and of experimental material from many different species of vertebrates. Nerve fibres that have been separated from their cell bodies by injuries or experimental cuts degenerate and the degenerating fragments can be traced through the brain by special stains. Between 1870 and 1950 most of the major pathways of the brain had been traced and in 1949 we were taught about these two distinct pathways that carried sensory messages up to the thalamus for relay to the cortex.

The information about the ascending spinothalamic and lemniscal sensory pathways is well founded and is widely taught. However, like so much of our knowledge about the brain, it is incomplete. For example, Ramón y Cajal (1955) had earlier illustrated axons that ascend through the spinal cord also giving off branches that feed directly or trans-synaptically[14] to motor branches and contribute to the spinal mechanisms that control walking and posture (Grillner 2003) (see Figure 2.6(a)).

These connections were not mentioned to us students in relation to the ascending pathways because the central problem was perception, not motor control. Motor control was a different lecture and is still a different chapter in most textbooks. Similarly, Ramón y Cajal's illustration of the **medial lemniscus** giving off branches to a **nucleus** in the brainstem called the **red nucleus** (Figure 2.6(b)), which in turn sends axons back to the spinal cord and plays a role in motor control, were not mentioned as a part of the ascending pathways. If we wanted to understand perception we needed to learn how the messages reach the cerebral cortex, and for this the pathways to the thalamus sufficed. The other connections were treated as irrelevant, or rather, not treated at all. Our teachers were, I am sure, not aware of them.

Comparably, during the nineteenth century, the visual pathways had been traced from the retina to the **lateral geniculate nucleus**, the thalamic relay for visual inputs to the cortex of the **occipital lobe**.

---

14 Involving a relay in more than one cell or cell group.

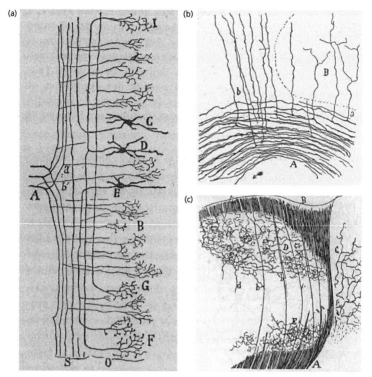

Figure 2.6 Representations of sensory nerve fibres that carry messages on the way to the thalamus and also give rise to branches that innervate centres concerned with motor control.

(a) Spinal dorsal root fibres (A) entering the spinal cord and sending ascending branches for relay to higher levels including the thalamus. Note the many branches that are given off at local levels (e.g. a, b) to innervate local spinal mechanisms (B, G, and F). The crossed fibres of the spinothalamic pathways (see Figure 2.5) are not shown in this figure.

Golgi preparation from Ramón y Cajal (1955).

(b) Fibres of the medial lemniscus on their way to the thalamus giving off branches to a nucleus in the midbrain, the red nucleus, which itself is known to have motor outputs to spinal levels.

Golgi preparation from Ramón y Cajal (1955).

(c) Fibres of the optic tract at the bottom (A) passing on their way towards the superior colliculus, which lies beyond the region marked B at the top of the figure. The individual axons give off branches to the lateral geniculate nucleus (shown at D and E) as they pass on their way to the superior colliculus in the midbrain.

Golgi preparation from Ramón y Cajal (1955).

That explained visual perception for the medical class. Observations that the lateral geniculate nucleus actually receives its visual inputs as branches from retinofugal axons as they pass from the eye to the superior colliculus in the midbrain (Figure 2.6(c); see also Figure 1.1(b) in Chapter 1) were not mentioned to our 1949 class. The superior colliculus, like the red nucleus, has direct and indirect motor outputs to brainstem and spinal motor mechanisms. When, later, the rich retinal innervation of the superior colliculus was once more a focus of interest, it was at first treated as a second visual pathway (Sprague 1966; Schneider 1969) for messages to be passed to the cortex, and the relationships between these separate visual centres became of interest (see Chapter 4 for discussion).

For the **auditory pathways**, we also learnt about the complex of brainstem relays to the medial geniculate nucleus of the thalamus for relay to the auditory area of the temporal cortex. We were not taught that the superior colliculus also receives auditory inputs that in the collicular locus of their terminations match visual inputs, in terms of their general directions, or that incoming axons from the auditory nerve at their entrance to the brainstem give off branches that descend towards the spinal cord (Harrison and Warr 1962). These are thought to underlie the so-called **startle response** to a loud handclap that is useful for testing a newborn baby's hearing (Lee et al. 1996).[15] The focus was on perception, on getting the messages to the cortex, so that the cortex could develop a reaction to the messages coming from the environment.

This view of the ascending pathways as 'pure' sensory inputs was still taught last year (2015) here at the University of Oxford when I helped with a practical neuroanatomy class for medical students. The idea that these pathways might better be regarded as carrying sensorimotor messages on the basis of which we can interact with the world, has remarkably little appeal. The strong belief in a real world out there created for us by a benevolent biblical deity had a strong hold on many

---

[15] They may also contribute to the difficulty we experience when we try not to walk in step with a marching band, or even a single drummer.

investigators during the second half of the nineteenth century when the pathways were being defined. Also today, it often seems difficult for Western neuroscientists to accept a world that depends on unconscious conclusions rather than one that we can directly and immediately access through our senses.

Chapter 3

The pathways for action

3.1 **Summary**

Early nineteenth-century studies demonstrated, on the basis of clinical, experimental, and anatomical evidence, that a motor pathway, the corticospinal or **pyramidal tract**, passes from a specific area of the cortex, the precentral motor cortex, to the brainstem and spinal cord. The **motor cortex** can be seen as a topographic map of the movable body parts, and damage to the cortex or pathways produces correspondingly localized paralysis. However, there are a great many other pathways that link other areas of the cortex to parts of the brain active in the control of movements. These still play a puzzling role in the standard model where the control of movements focuses on cortical contributions to voluntary movements by the corticospinal pathways.

3.2 **Early studies**

Our view of the motor pathways also has a long and imposing history. It is worth looking at some of this history briefly to understand a view of cortical functions that puts the cortex in direct control of actions through its connections with the spinal cord and brainstem, without involving the intermediate motor centres even though, as outlined in Chapter 4, these centres receive rich cortical inputs and have a long evolutionary history of controlling complex movements in our non-mammalian ancestors who lacked a six-layered cortex.

One important point about understanding the motor pathways concerns the localization of function in the motor system: today we know that different parts of the motor cortex connect to different parts of the body, just as on the sensory side different parts of the body connect to different parts of the post-central sensory cortex (see Figure 2.5).

This general concept of localization of function has, however, had a chequered early history. Although both Descartes (see Figure 2.1) and Newton had proposed that the visual world is represented centrally by an ordered map (as shown by Figure 8.1), there followed a period of more than 100 years during which localization of functions in the brain was, first, firmly denied by Fluorens and his allies and, a little later, extravagantly claimed by the **phrenologists** Gall and Spurzheim and their many, fashionable, enthusiastic followers (see Finger 1994).[16]

Three important studies in the second half of the nineteenth century demonstrated an area of cortex concerned with the control of movements and showed that different parts of the cortex relate specifically to the movements of different body parts.

Hughlings Jackson (1881) studied the brains of patients who had suffered from **grand mal epilepsy**. He noted that for any one patient the seizures would start at either the fingers, the toes, or the mouth. These are the parts of the body having the densest motor and sensory innervation. Post-mortem studies showed that in the brains of these patients, cortical lesions were localized to the **precentral gyrus** (labelled 2 in Figure 1.1(a)), and for a patient whose seizures started in the face, the lesion would be in the lowest parts of this gyrus, whereas lesions in the upper parts of the gyrus would relate to seizures starting at the toes. Lesions between these two would relate to seizures starting in the fingers. This led Hughlings Jackson to conclude, correctly, that these separate parts of the precentral cortex were concerned with the control of movements in the related body part.

In 1870, Fritsch and Hitzig had identified, in the brains of dogs, a specific cortical area stimulation of which produced movements that involve different body parts depending on the precise site of the stimulus, confirming the view of a topographically organized motor cortex (Fritsch and Hitzig 1870). A few years earlier, **Broca** (1863) had reported that damage to a region of the **frontal lobe** just above the fissure that separates it from the anterior part of the temporal lobe

---

[16] The views against any functional localization in the brain were comparable to the holistic views of Golgi, mentioned in Chapter 5.

(labelled 4 in Figure 1.1(a)), was associated with motor **aphasia**; that is, an inability to produce speech even while comprehension of speech survived.

Clinical studies of patients who had suffered lesions of the precentral motor cortex (labelled 2 in Figure 1.1(a)), or of the axons that pass from the cortex through the **internal capsule** (labelled 5 and 6 in Figure 1.1(d)), showed severe motor losses on the side contralateral to the damage. The patient was unable to initiate movements in the affected body parts even though the clinician could still elicit reflex activity from them. This confirmed the view of the important role in motor control of the precentral motor cortex and its outputs.

Once the motor cortex had been identified, its contribution to the important **corticospinal tract** could be established. Post-mortem studies of patients with motor losses due to cortical lesions showed that the nerve fibres in this tract degenerate, and the degenerating fragments could be traced back to a lesion in the internal capsule or motor cortex. The fibres pass from the cortex through the internal capsule (labelled 5 and 6 in Figure 1.1(d)) and from there through the brainstem, passing through the **cerebral peduncle** and the **pons**, to emerge from the **caudal** parts of the corticospinal tract, as a compact fibre bundle, the pyramidal tract[17] to the spinal cord with a crossing just before they enter the spinal cord. They end in relation to the nerve cells in the spinal cord, including those that directly innervate the muscles.

Also, following the early observations that defined the motor regions of the cortex, other studies, first by Grünbaum and Sherrington in the gorilla, chimpanzee, and orangutan (Grünbaum and Sherrington 1901), later by Penfield and others in humans (Penfield and Boldrey 1937),[18] confirmed that electrical stimulation of particular cortical regions could produce movements in particular body parts, and demonstrated the details of the layout of the precentral gyrus in primates postulated

---

[17] Forming the swellings called the pyramids (labelled 8 in Figure 1.1(c)).

[18] They were applying the stimuli to the cortex in conscious patients in order to define an epileptic focus so that Penfield could limit the necessary cortical removal to the precise region that was responsible for the epileptic discharges.

by Hughlings Jackson. Campbell (1905) produced a strikingly detailed study of functional localization in the motor cortex: he studied post-mortem brains of many patients who had earlier had whole limbs or limb parts amputated. Although the motor cortex innervates the spinal motor cells, so that the amputations had not affected the corticospinal cells directly, the motor axons from the spinal cord had their axons cut, and many of those cells had died or were shrunken. The cortical cells showed what has been called a transneuronal retrograde degeneration in cortical zones that closely matched the localization postulated by Hughlings Jackson for the human brain and demonstrated by Grünbaum and Sherrington for the anthropoid apes not just for whole limbs but for limb parts.

The increasing knowledge of cortical mechanisms of movement initiation, either of the simple movements produced by stimulation of the motor cortex or the much more complex and sophisticated movements coming from Broca's motor speech area, led to a view of motor cortical regions as generators of voluntary movements and in particular to a search for further information about the connections of the precentral motor cortex. These included studies of the details of the outputs of the motor cortex to the spinal motor apparatus (Landgren et al. 1962; Evarts and Tanji 1976; Georgopoulos et al. 1982) as well as studies of the inputs to the motor area from possibly 'higher' pre-motor areas (e.g. Passingham et al. 2010). The crucial issue for the motor pathways as for the sensory pathways is that they are topographically organized in relation to the body parts.

3.3 **Some more recent observations**

I have briefly summarized the important evidence from lesions and from cortical stimulation, which demonstrates a major role for the motor cortex and the corticospinal tract in the production of voluntary movements, in order first to stress the importance of this evidence for the standard view and then to compare it with observations of monkeys made some years ago by Lawrence and Kuypers (1968a, 1968b). These authors showed that in monkeys, a lesion of the motor cortex or of its output axons in the cerebral hemispheres produces severe paralysis on the opposite side of the body, comparable to that seen in patients. However, in contrast to this, a selective interruption in a monkey of the corticospinal pathway just after it leaves the region of the pons (labelled 6 in Figure 1.1(c)), where the corticospinal tract forms the **pyramids** (labelled 8 in Figure 1.1(c)) just before the fibres cross the midline, has a different and unexpected effect. After a short recovery period, the monkeys are able to move about nimbly and climb about their cages. The major demonstrable motor loss concerns finer finger movements like those that a normal monkey uses to get food out of a small food container specially designed to demonstrate the animal's ability to pick out small bits of food between finger and thumb. After such a lesion, the monkey needs to scoop the morsel out with the whole hand.

The important point made by these experiments is that the spinal portion of the output fibres of the motor cortex, that is, the corticospinal axons cut just caudal to the pons in these experiments, represents just one part of the total output, direct and indirect (see Figure 2.2), from the cortex to the spinal motor outputs. It does not include the many motor outputs that travel more laterally, in the brainstem, not in the pyramidal tract, and which have relays in subcortical centres (e.g. Holstege and Kuypers 1987; Holstege et al. 1988). These lateral pathways involve relays in nuclei of the brainstem, including the red nucleus and the midbrain locomotor area mentioned earlier (see Chapters 1 and 2). The red nucleus, in addition to the relatively small input coming from the medial lemniscus (Figure 2.6(b)), receives direct inputs from the cerebral cortex and from the cerebellum and sends descending motor outputs to the spinal cord. These lateral pathways also include the superior colliculus in the midbrain (labelled 8 in Figure 1.1(b)), which sends outputs to the motor centres of the brainstem and through them to the spinal cord and itself receives rich inputs from the cortex (Harting et al. 1992; and see Cerkevich et al. 2014). It is a phylogenetically old centre, which contributes to the control of gaze, that is, of head position and eye position (Kardamakis et al. 2015). John Harting and colleagues demonstrated for cats, 25 different cortical areas that send direct projections to the superior colliculus (Harting et al. 1992). I once asked John whether he had ever found a cortical area that did not project to the colliculus and he said he had not.

Perhaps most importantly, the cortex also sends extensive inputs related to motor control to the basal ganglia (labelled 1–3 in Figure 1.1(d)) and also through the pons to the cerebellum. These pathways, all apparently concerned with providing adjustments of the final movements, are spared by the lesion of the corticospinal tract caudal to the pons. Many cortical areas can continue to contribute to a greater or lesser extent to what a monkey can do even without the direct corticospinal pathway. The significant 'extrapyramidal' pathways that include most cortical areas and link them to the lower motor centres can help us to understand the observations reported by Lawrence and Kuypers (1968a, 1968b). They demonstrate that these pathways control significantly more than the 'reflexes and postural adjustments' suggested by the right side of Figure 2.2 and that they can suffice to support the phylogenetically old brainstem and spinal mechanisms in producing adequate movements.

In summary, the evidence shows that the pathways needed for the standard view do indeed exist and do have some of the functions proposed by the standard view. However, there is clear evidence that these pathways form only a part of a complex set of connections in mammalian brains. These involve many of the phylogenetically older subcortical regions. They controlled movements during the long evolutionary history of vertebrates and can still make significant contributions in mammals.

Chapter 4

The subcortical motor centres

4.1 Summary

This chapter looks more closely at some of the subcortical motor centres that often play a peripheral, or an auxiliary role in the standard view: primarily the basal ganglia, the cerebellum and the superior colliculus; also several brainstem centres. These all play a significant role in motor control and between them receive inputs from the majority of cortical areas. The colliculus serves as an example of a centre that in mammals is often dominated by the cortex. The cortical action may be direct or may involve a strong inhibitory pathway through the basal ganglia. The standard view assigns even quite simple actions to the motor cortex, although comparable actions can be controlled in our vertebrate ancestors by the midbrain **tectum** which corresponds to the mammalian **superior** and **inferior colliculi**. The interactive view has information about movements going to most parts of the cortex, and has all cortical areas contributing to motor control through phylogenetically old centres. For most cortical areas, we must still learn how their motor outputs influence our actions.

4.2 A brief overview of some of the major subcortical motor centres

Some of the major subcortical centres involved in the control of movements mentioned at the end of Chapter 3 are shown in Figure 4.1. These centres all have access to brainstem and spinal motor outputs, they all depend significantly on direct or indirect inputs from the cortex, and they all played a role in the control of movement in our non-mammalian vertebrate ancestors, who lacked a hierarchy of six-layered cortical areas with their complex thalamic links, but nevertheless, managed to produce remarkably accurate, skilful movements. In mammals, these centres all receive direct cortical inputs, often from many cortical areas. We have to think about the extent to which the phylogenetically old parts of our brains might still play a large role in the production of movements, to ask whether the newly acquired mammalian cortex replaces the old machinery or exploits it.

Three of the centres shown in Figure 4.1 are of particular relevance. These are the basal ganglia (BG), the superior colliculus (SC), and the cells of the pons (Po), which link cortex to the cerebellum (from pathways 5 and 13 in Figure 4.1). Each of these centres receives inputs from a very large number of cortical areas (represented in the figure by 1, 4, and the several pathways included together in 5). The many different cortical areas that provide corticofugal inputs to these three areas leave essentially no part of the cortex without a connection to one or more of these phylogenetically old, subcortical motor centres.

We need to define not only what it is that a mammal can do with its large, extensive cerebral cortex that our vertebrate ancestors could not do (this is considered further in Chapter 13), but also to learn to what extent and under what circumstances any one of the several subcortical centres functions under cortical control or runs freely, independent of cortex. These are questions about how we use the cortex, and how we use our older motor centres. Perhaps the cortex is most important for learning new skills, and once we have learnt a skill, we function like our ancestors without cortical control (see Grillner 2015). These questions about cortical functions will force us to think about how the subcortical parts of the brain relate to the cortex and how our behavioural skills depend on the cortical and subcortical centres.

In the next section I will focus on the superior colliculus, using this as an example of subcortical functions, but that should not be taken to imply that other lower motor centres illustrated in Figure 4.1, which also receive inputs from thalamocortical circuits, are likely to be less interesting or important once their functional links from cortex and to other centres are understood.

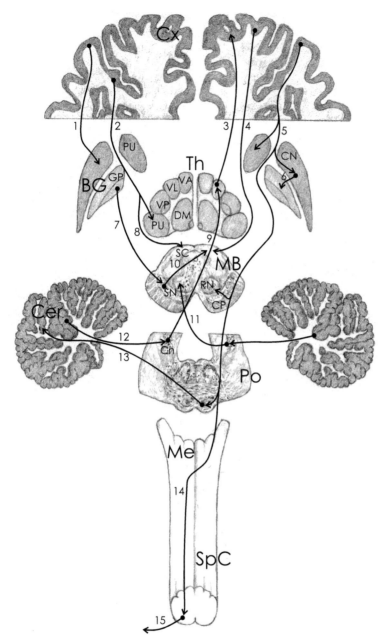

Figure 4.1 Summary figure of some of the connections discussed in the text that link the subcortical, phylogenetically old motor centres of the brainstem and spinal cord to the thalamus and cortex. The images of the cortex (C), basal ganglia (BG), and thalamus are taken from Figure 1.1(d) in Chapter 1.

Large labels: BG, basal ganglia (includes putamen (PU), caudate nucleus (CN), and globus pallidus (GP)); Cer, cerebellum; Cx, cerebral cortex; MB, midbrain; Me, medulla; Po, pons; SpC, spinal cord; Th, thalamus.

Small labels: Cn, deep cerebellar nuclei; CP, cerebral peduncle; DM, dorsomedial thalamic nucleus; Pu, pulvinar; RN, red nucleus; SC, superior colliculus; SN, substantia nigra; VA, ventral anterior nucleus; VL, ventral lateral nucleus; VP, ventral posterior nucleus.

Illustration by Dr Lizzie Burns.

In particular, although I will not consider this in detail, it is important to stress that the cortex through its connections with the basal ganglia plays a role in the initiation, the cessation, and the strength of movements, and through its connections through the pons with the cerebellum plays a role in the timing and the accuracy of movements. These are oversimplifications, but they stress that without these two major regions, both present in the old vertebrate brains, movements would be much less efficient. Both receive rich inputs from many cortical areas. All parts of the cortex are in a position to contribute to the control of ongoing movements through these pathways. This is not surprising according to the interactive view, where every cortical area receives information about forthcoming movements.

4.3 Can the midbrain tectum control actions that do not involve the cortex?

The roof of the midbrain, the tectum, formed in mammals by the superior colliculi (labelled SC in Figure 4.1) and inferior colliculi (see Chapter 1, Figure 1.1(d)), is a single structure in non-mammalian vertebrates, there simply called the tectum (roof) of the midbrain. The midbrain tectum played a significant role in the control of movements in our vertebrate ancestors and in mammals it illustrates some of the complex interactions that link thalamus and cortex to phylogenetically older centres.

In mammalian brains, the superior colliculus still has rich two-way connections with spinal and brainstem nerve roots for incoming and outgoing messages (not shown in Figure 4.1). It has important brainstem outputs for the control of eye movements and also has outputs to the spinal cord that control head movements in coordination with eye movements. It sends inputs to the thalamus (Huerta and Harting 1982; Bickford et al. 2015) and also receives inputs from a great many cortical areas (Harting et al. 1992).

In fish, frogs, reptiles, and mammals, the tectum is a multisensory structure (Stein and Arigbede 1972; Dräger and Hubel 1975; Stein and Gaither 1981; Deeg and Aizenman 2011; Yu et al. 2014), receiving somatosensory,[19] visual, and auditory inputs that are mapped in register with each other in terms of major directions. The non-mammalian tectum itself has the capacity to organize the major actions needed, for example, by a fish or a frog to catch a fly in water or air. Rácz et al. (2008) showed the tectal pathways that link visual inputs to the control of movements of the tongue. These pathways are likely to play a significant role in the frog's capacity to catch flies. Comparably, there is recent evidence that the superior colliculus of primates has the neural connections needed for directing arm movements appropriately towards objects in space (Philipp and Hoffman 2014). We need to understand the extensive cortical inputs to the tectum in relation to what the tectum can do on its own.

There are good reasons for thinking that cortical activity often dominates the activity in the superior colliculus[20] and this relationship is well illustrated by experiments summarized in Figure 4.2, on the visual capacities of cats: whereas a cat with all parts of its **visual cortex** destroyed on one side loses its capacity to orient towards visual stimuli (small food rewards) presented on the side opposite to the cortical damage[21] (see Figure 4.2 (left)), the same cat with the contralateral superior colliculus removed (Figure 4.2 (middle)) or the tectal commissure (a bundle of fibres that connects the colliculi across the midline) cut (Figure 4.2 (right)) recovers the ability to orient (Wallace et al. 1990). It appears that at least a part of this recovery is due to interruption of the crossed pathway (labelled 10 in Figure 4.1)) from the pars reticularis of the substantia nigra (labelled SN in

---

[19] Somatosensory refers to a group of sensory pathways from the body, the limbs, and the head that include messages for touch, pain temperature, and deep pressure as well as messages from joints, ligaments, and muscles, providing information about limb movements.

[20] For example, the observation that response properties of collicular cells in cats are normally dominated by cortical inputs but are more like the response properties of retinal ganglion cells when the cortex is silenced (see Wickelgren and Sterling 1969) is relevant.

[21] The cortex receives its visual inputs from the opposite side.

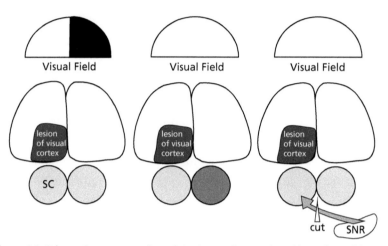

Figure 4.2 Schematic representation of the interaction produced by unilateral lesions of the visual cortex and the superior colliculus, based on the experiments described by Sprague (1966) and by Wallace et al. (1990). Whereas a lesion of the left visual cortex produces a loss of visual reactions for the right visual field (left), a lesion of the right superior colliculus (middle) or of nigro-collicular fibres that cross the midline between the two colliculi (right) leads to a recovery of visual reactions lost due to the cortical lesion.

SNR, substantia nigra pars reticularis.

Figure 4.1),[22] to the opposite colliculus. These are inhibitory fibres, that is, fibres whose activity tends to reduce activity of the recipient collicular cells.

This inhibition represents the output of the basal ganglia (labelled 7 and 10 in Figure 4.1). The basal ganglia play a crucial role in the control of movements. They have inhibitory connections to many other brainstem regions and also to the thalamic nucleus (VA in Figure 4.1), which provides thalamic inputs to the motor cortex (not shown in Figure 4.1). The major outputs of the basal ganglia are inhibitory outputs to the thalamic nuclei concerned with movement control, and also to motor centres of the midbrain, including the pathways 7 and 10 shown

---

[22] The pars reticularis of the substantia nigra is a group of cells in the midbrain lying just above the **cerebral peduncles**. These cells receive significant inhibitory inputs from the basal ganglia, and themselves send inhibitory fibres to the thalamus, the superior colliculus, and other regions of the midbrain.

in Figure 4.1. The inputs to the basal ganglia, from which pathways 7 and 10 arise, come from widespread areas of cortex and from several thalamic nuclei, providing links with the world, the body, and other brain regions.

This simplified account is sufficient to indicate that the fibres that pass from the **substantia nigra pars reticularis** to the superior colliculus provide indirect inhibitory controls from cortex to the superior colliculus. In addition to this are the many more direct, and functionally mostly undefined controls that come directly from the cortex to the colliculus.

These observations suggest that the cortex and thalamus acting through the basal ganglia can inhibit movements that are controlled by the colliculus. The cortex acting through the basal ganglia (labelled 1, 7, and 10 in Figure 4.1) can close down the collicular control of movements, presumably so that the cortex can take over, and this implies that there are conditions when the colliculus can itself be in control. The extent to which higher centres, particularly cortex and **striatum**, determine the level at which actions are, or are not, initiated and controlled by lower centres merits detailed studies of the conditions under which these pathways are active.

Philipp and Hoffmann's recent (2014) demonstration that electrical microstimulation of the deep layers of the superior colliculus of macaque monkeys can produce 'more or less stereotypical arm movements' suggests that in primates, as in fish and frogs, the circuitry needed for producing movements oriented to a moving object exists in the colliculus. However, the extent to which this capacity is initiated by the cortex or is normally held in check by inhibitory pathways from the substantia nigra, under instruction (presumably) from the cortex and basal ganglia, is not known. We need to learn more about the cortical cells and the nigral cells that may be innervating the tectal cells from which Philipp and Hoffman produced arm movements. For now, the primary conclusion is that phylogenetically old parts of mammalian brains have circuitry that may well be capable of producing well-oriented and well-timed movements of the limbs or the eyes.

This conclusion applies to the spinal mechanisms that can control scratching or walking (Sherrington 1906; Grillner 2003; Grillner et al. 2007), and to the pathways that are contributing to the complex cage-climbing behaviour described by Lawrence and Kuypers (1968a, 1968b) after lesions of the corticospinal tract. It also relates to the cerebellum and to the brainstem motor centres receiving cerebellar outputs (not shown in Figure 4.1),[23] as well as the midbrain locomotor centre considered in Chapter 1 and to the tectal capacities that have been discussed in this section.[24] The possibility that these older centres have lost their capacities for contributing to motor control and that all movements must be initiated from the cortex, is something that is seriously entertained by many investigators on the basis of the severe losses that follow cortical lesions. However, as demonstrated by the Sprague effect (Figure 4.2), this may be an error because the lesions have more complex actions than merely the loss of the direct cortical output from the specific damaged cortical area to the brainstem and spinal motor centres.

There is a high probability that many actions in a normal mammal are produced in response to stimuli that do not reach the cortex and thus are probably not, in themselves, conscious actions. They may be a part of a complex sequence of actions that can be brought under conscious control, such as a return in a tennis game, or touching the footbrake while driving a car, but may themselves, once well learnt, be largely or

---

[23] Particularly the red nucleus, which also receives cortical input and sends outputs to the spinal cord, but also other less clearly defined brainstem motor centres.

[24] The phenomenon of 'blindsight' (see Humphrey and Weiskrantz 1967; Weiskrantz et al. 1974) defined as the ability of patients (or monkeys) with visual losses due to lesions of visual cortex, to point at visual stimuli in the blind parts of their visual fields, which the patients claim (convincingly!) not to see, has been studied in great detail and has often been ascribed to preservation of visual inputs to higher visual cortical areas. The observation (Sahraie et al. 2013, page 18337) that: 'In the majority of cases (17 of 20), the presence or absence of significant detection abilities [i.e. the capacity to respond to the visual stimulus] determined psychophysically matched the presence or absence of a significant pupil response' suggests that a significant portion of the blindsight demonstrable after cortical lesions may depend on the midbrain, which receives not only the rich inputs from the cortex but also receives direct retinal inputs many of which have also sent branches to the lateral geniculate nucleus.

entirely subcortical actions, like the walking sequences of a cat with an isolated spinal cord. Many non-cortical movements can probably be produced, like the scratch reflex demonstrated by Sherrington in 1906, accurately and efficiently by the lower, spinal and brainstem, centres. We need to learn much more about what the lower motor centres can do in an intact brain, how they interact with each other, and to what extent their functions can be modified or completely inhibited by activity coming from the cortex. We can think of the phylogenetically old centres as an old horse that can find its way from home to the village pub and back even without the farmer or with a drunk or sleeping farmer. The same old horse will undertake more varied and interesting trips when the farmer is present, awake, and sober.

4.4 **Another look at the role of the cortex in the control of movements**

I have provided in Chapter 3 a brief view of how, in the standard view, peripheral sensory inputs pass to the primary somatosensory cortex. In terms of further processing in the cortex (for perception or action), these sensory messages are passed to higher areas of the **parietal cortex** (summarized by Andersen (1995) and see Buneo and Andersen (2006)) and then to frontal cortex for passage from the premotor to motor cortex. Churchland (2002) presents a clear view, based on Andersen's account, of what may be happening in the parietal cortex (see Figure 4.3). She presents a theoretical treatment of the complex computations that the brain needs to undertake in order to match coordinates of the sensory inputs from the eyes or the ears to the coordinates of the motor apparatus. She describes someone who hears the whine of a mosquito, sees the mosquito, and then 'effortlessly' slaps the mosquito. She writes (page 310): 'The effortlessness makes the task seem easy, but computationally it is anything but simple. The central point is that sensory coordinates have to be transformed into motor coordinates in order to connect with a sensorily specified target'.[25] She

---

[25] The Cartesian separation of sensory from motor pathways here is crucial to the conclusions.

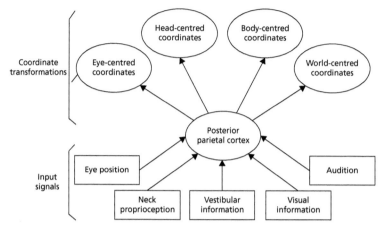

Figure 4.3 This figure illustrates the connections that may be needed in the parietal cortex in order for a person to be able to swat a mosquito successfully. The figure shows how the sensory inputs at the bottom of the figure need to be converted into appropriate coordinate systems, shown at the top of the figure, so that an accurate motor output can be produced.

Reproduced from Patricia Churchland, *Brain-Wise: Studies in Neurophilosophy*, Figure 7.17, © 2002 Massachusetts Institute of Technology, by permission of The MIT Press.

illustrates this by a schema that has a hypothetical area of the parietal cortex receiving five different sensory inputs (see Figure 4.3) and, in turn, connecting to four different centres responsible for the necessary coordinate transformations. She postulates an 'arch mapper' that can integrate the necessary information and generate an accurate movement from the relevant sensory inputs. This arch mapper 'cannot neatly be described in everyday terms' but from her description it has many of the attributes one might look for in a mind.[26]

It looks as though this need for an arch mapper, and the involvement of the parietal cortex in an action no more complex than that of a frog catching a fly, may be a necessary part of a complex theoretical model specially created for mammals in which the many sensory-to-motor connections are all categorized as depending on the cortex as in Figure 4.3. This view of the mammalian brain as depending on the

---

[26] Provided, of course, that one thought that a mind plays a part in the control of neural circuits, a view that Churchland does not support.

cortex for even quite simple actions is one that has to be challenged. In non-mammalian vertebrates, such as frogs, lizards, or fish, subcortical centres control sensorimotor integration, and these centres are still present in mammals. Visual, auditory, and **somatosensory** information reaches not only the cerebral cortex but also reaches the mammalian superior colliculus along pathways that also exist in frogs and lizards. As indicated earlier, in the mammalian colliculus these three sensory maps for vision, hearing, and touch are roughly in register in terms of major directions. That is, there is a midbrain structure that has all of the sensorimotor visual, auditory, and tactile inputs that are needed for swatting mosquitoes or catching flies. There are pathways from the superior colliculus to the spinal cord, some with and some without a relay in the brainstem reticular formation. The control of arm movements that Philipp and Hoffmann (2014) demonstrated in monkeys by stimulating the deep layers of the colliculus, indicate that, yes, indeed, pathways needed for swatting flies are likely to be present in the mammalian superior colliculus.

More significantly perhaps, we have seen that the messages carried from spinal levels up to higher levels include information about sensory inputs and also about the immediately related motor responses. They are *sensorimotor* from the first branches formed soon after they enter the central nervous system. No matter where they go, they are carrying sensorimotor information. The coordinates of the sensory information are already linked to the coordinates of the related, forthcoming actions in each of the relevant sensory pathways. The role of the parietal cortex in swatting flies may turn out to be relatively minor.

This is not to trivialize the integrative functions of the parietal cortex. Rather, it is to stress that the parietal cortex is likely to be doing something more interesting and more complex than is needed for swatting mosquitoes. It is highly improbable that the characteristic thalamocortical complex of mammals evolved merely in order to do something that our non-mammalian vertebrate ancestors already did efficiently, largely by depending on their midbrains, cerebellums, and basal ganglia. It seems more likely that the parietal cortex may be capable of doing something altogether different from the computation proposed in

Churchland's model. Possibly, at the beginning of a mosquito-infested vacation, one has to learn to let the colliculus and lower centres take charge of the actions, and the cortex may play a role in that (Grillner 2015; Kawai et al. 2015), but someone who has spent a long, hot summer among the lakes of Minnesota may well be able to swat mosquitos without any cortical involvement at all, using the midbrain connections as do frogs when catching flies. It should be possible to define the neural links that play a role in such interactions in primates rather as Rácz et al. (2008) traced the pathways from the visual inputs to the tongue muscles of the frog and demonstrated a plausible midbrain fly-catching circuit. A good place to start would be with the cells in the deep layers of the superior colliculus of monkeys from stimulation of which Philipp and Hoffmann (2014) produced arm movements. Such studies could then be followed by investigations of the functional links that are established as the action is first learnt and then established as an efficient and rapid reaction to the sound, the sight, or the bite of a mosquito. The cortex may then play a role in our perception of the act but not in its rapid execution.

For much of contemporary neuroscience, the cortex is often treated as though it is an obligatory part of actions like swatting mosquitoes, without asking what it is that the cortex can do that the midbrain cannot do. It may well be that perceptual processing[27] may need the cortex whereas sensory processing without conscious perception can be done at subcortical levels (see Chapters 12 and 13). Further, we have seen that the superior colliculus may often be dominated by, or even controlled by, cortex or basal ganglia. It is likely that many routine activities depend on the extent to which this cortical dominance of lower motor centres can be inhibited through the basal ganglia.

The separate pathways of the standard view, sensory into the brain for perception and motor out of the brain for action, certainly exist. However, we have seen that the problem is not their existence but their relevance for many of our actions. These pathways are an abstraction

---

[27] Perception here refers to a conscious process whereas sensation refers to a process that is not consciously perceived. The distinction is discussed further in Chapter 12.

from a far more complicated set of interconnected subcortical parts, many of which played important roles in the lives of our non-mammalian ancestors and all of which are still relevant for us as mammals in our daily lives. Only a few are shown in Figure 4.1. The rich, complex, but still poorly understood nature of the subcortical pathways does not in itself prove the standard view wrong. It suggests that other possibilities are worth exploring.

We are now in a position to ask the question that must arise from any comparative study of vertebrate brains: what is it that the thalamus and cortex together provide for mammals that is not present in their vertebrate ancestors?

Part II

My route to the thalamic gate

Chapter 5

Starting to study the brain

5.1 Summary

Chapter 5 takes a look at some of the basic assumptions, questions, rules, laws, and dogmas that we encounter, use, or ignore when we study the brain. As a student I was fortunate to have been taught to see the brain as a functional part of a whole animal, and the nervous system as a part of biology. I was introduced to the 'neuron doctrine', which was applied specifically to studies of the brain. I became aware of its strengths and gradually learnt its weaknesses. I learnt the difference between descriptive science and science as a hypothetico-deductive system, and began to see myself as a descriptive scientist looking for areas of the unknown still needing to be explored. I have used this chapter to illustrate, that for the most complex parts of the brain, the rules that are useful where we know and understand many details, are often irrelevant or confused where, as for the thalamus and cortex, we know only a few of the relevant details. This book has been written about areas where we are trying to understand neural pathways and their role in behaviour and cognition, where we often lack sufficient details, and where crucial questions have not been asked because they did not arise under the standard perception-to-action model. New questions become relevant under the interactive view, providing opportunities for many new, needed descriptive studies.

5.2 Learning to think about the functions of the brain

My PhD studies in the Anatomy Department at University College London from 1951 to 1953 dealt with a small part of the brain just below the thalamus, called the **hypothalamus**. It had recently been the subject of a number of publications (e.g. Bard and Mountcastle 1948; Ranson 1936) describing how this part of the brain controls important 'visceral' reactions such as blood pressure, heart rate, gut motility, and respiratory rate. R Hess (Hess et al. 1950: Akert 1981) had recently won a Nobel Prize for his studies of the hypothalamus in cats and WE Le Gros Clark, the Professor of Anatomy at Oxford at the time, whom I had known briefly when I was 12, had edited a book on the hypothalamus. My thesis supervisor, JZ Young, had suggested a thesis on the cerebral cortex of birds,[28] but he wisely and generously gave me complete freedom to choose the hypothalamus instead, although he quite clearly had no interest in this part of the brain himself.

JZ Young played a very significant role in my education and in my choice of neuroanatomy as a career, in a way that is difficult to summarize. His influence extended far beyond the freedom he gave me to follow my own inclinations in my research. My route to a strong interest in the thalamus and its relationship to the cortex was, as the following account shows, long, slow, and rather devious, and to a very significant extent depended on luck. However, it also depended on my choice of the hypothalamus, which has a small specialized region in mammals, the **mamillary bodies**, that send an important link to cortex through the thalamus. It was the mamillary bodies that led me to the thalamus and cortex.

My progress also depended on what Professor Young had taught us early in our introduction to anatomy. He taught us to think about the parts of the body not in terms of memorized details, but in terms of broad brush strokes needed for learning about basic biological issues. What might any one part of the body be contributing to the life of the animal? That is, he taught us to think, while we were studying their parts, about what animals do, how they live, and how, in their natural lives, they interact with the world. In 1950 he published a book, *The Life of the Vertebrates*. It was not a book about vertebrate anatomy

---

[28] JZ Young had recently moved to University College London from Oxford, where Tinbergen had been undertaking important studies on the development of behaviour in birds. Perhaps this had been a great missed opportunity for me.

or physiology, or one on vertebrate evolution, but it was deliberately framed around the *life* of vertebrate animals. In the department at that time were two artists, Miss Turlington and Miss de Vere, who provided the many detailed illustrations for this book. Some were obtained from actual dissections or from observations of animals in the zoo (see Figure 5.1(a)) and even the pictures of skeletons (see Figure 5.1(b)) communicated a sense of a living animal 'within' the bones.

The lectures he gave to the entering medical class were original and focused on the active lives that had existed within the human bodies that the students were dissecting. The lectures were not concerned with collections of facts about the structures of those bodies, with the long lists that traditional anatomists valued. JZ (he was always known to us students as JZ, and I will use this in the following text) encouraged us to think about the structures we would be dissecting, not telling us about the several named parts of a bone but describing how the shape of a bone relates to stresses it is exposed to in life. He told us that the shape of a bone carried a 'memory' of those stresses. The idea that bones have memories was novel and strange. I was strongly reminded of it quite recently when I heard an Oxford philosopher, lecturing to a group of young neuroscientists, tell that the Nobel Prize that had been awarded to Eric Kandel for his studies of memory in the sea hare (or

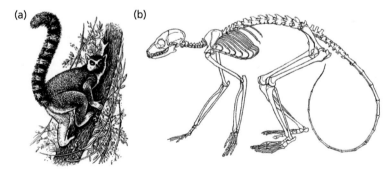

Figure 5.1 Two illustrations from JZ Young's 1950 book, *The Life of the Vertebrates*. (a) A Ring-tailed lemur and (b) its skeleton.

sea slug, *Aplysia*) was an error[29] because sea hares, by definition, have no memory. He based his argument on the Greek roots of the word. For JZ it was the interaction of the bones with their environment that was of interest because the environment left evidence about its actions in the structures of bones; this evidence can be read by the investigator to recreate events during the developmental and evolutionary history of the animals—a record of past events in the animals' lives and their ancestors' lives.

JZ spent a large part of his career studying memory in the octopus. Here was an animal with a brain that differs profoundly from the brains of mammals, and that yet was able to do some strikingly 'intelligent' things with its, to us, unusual brain. What was it in each of the two brains, ours and those of the octopus, that allows each in its own way to remember the items (coloured triangles, squares, circles, etc.) that JZ and his students used as stimuli in their training experiments with the octopus. I never joined the group of people that were studying the octopus with JZ, but I learnt the importance of a comparative view.

One can argue, with hindsight, about the degree to which the comparative approach was successful in actually demonstrating the shared features that might underlie the formation of memories in an octopus or a mammal. Whereas JZ was looking at the connectivity patterns between nerve cells that often even today need more detailed descriptions, and do not provide clear answers relating to the behavioural level, Kandel, working later, was looking at the level of the molecules in relation to cellular activities, where the answers are now often clearer. Both approaches were creative and logical. Where we face the problems of the unknown we need to risk new ways of asking questions and we need to be prepared for failures. It was my interest in a comparative approach that eventually led me to the thalamus and cortex.

One of JZ's most lasting contributions to neuroscience was his early description of the 'giant axons' found in the nervous system of the squid.

---

[29] This was the speaker's error: the Nobel citation for the three Nobel Laureates at the time, Carlsson, Greengard, and Kandel, was for 'discoveries concerning signal transduction in the nervous system'. Kandel's prize was specifically for demonstrating synaptic plasticity in *Aplysia*.

These are unusually thick nerve fibres that produce rapid responses,[30] allowing the squid to escape quickly from threats. JZ had described these axons in 1936, showing that they were formed by the fusion of many separate, thinner axons that formed the larger single axon (Young 1936). At that time these observations were taken by some to be contrary to the then powerful 'neuron doctrine', which played an important role in neuroscience between the 1890s and the 1970s, a theory which stressed that the processes of nerve cells never form fusions. It is described in the next section.

5.3 Theories, doctrines, and laws for the nervous system

It is important to take a look at some of the theoretical underpinnings of neuroscience. Today, I see neuroscience as a part of biology and that is my focus throughout this book. For others, neuroscience is rooted in psychology, reaches across to philosophy, or needs to rely on computer models. In Parts I to III, I treat the brain and its cells as a part of zoology, studied by anatomists, physiologists, pharmacologists, developmental biologists, pathologists, immunologists, geneticists, molecular biologists, and chemists. These studies share their strongest theoretical constructs with biologists generally: the theory of evolution, genetics and immunology, as well as molecular, cell, and developmental biology, and the related most important part of the cell theory—that cells can only be generated from cells. We still need to learn how the complexities of our mental lives, the concepts that psychologists have defined and philosophers have discussed, can make sense in terms of the known activities of nerve cells. The problems of fitting the rather large concepts of psychologists and philosophers into the tight trousers, made to fit the sparse and less generous conceptual structures currently available for understanding large groups of nerve cells and their functions, still remain to be solved. As others have argued (e.g. Churchland and Sejnowski 1992; Stevens 1994), a strong theoretical

---

[30] The greater the diameter of an axon, the more rapidly it conducts action potentials.

basis for understanding higher cortical functions is currently lacking. Churchland and Sejnowski (1992) have described neuroscience as data rich and theory poor, and they add that many areas of neuroscience are, nevertheless, still data poor in terms of recognizably relevant data. I cannot claim grand theories about cortex and thalamus in the rest of this book, but I can suggest some promising stepping stones and some new views towards a better understanding of the thalamus and cortex, where data are poor and many new observations are still needed. As will become apparent in later chapters, one current lack is the absence of detailed information concerning the specific structures and functions of many essential neural connections in the brain. We still need much plain, descriptive science.

5.3.1 The neuron doctrine

The neuron doctrine dominated neuroscience between 1890 and 1970. It described the nerve cell as the basic developmental, structural, and functional unit of nervous systems, as well as the trophic unit in reactions to injury, and it fiercely denied that nerve cells or their processes could fuse to be continuous one with another. Although it was often claimed as a neural equivalent of or match for the cell theory, seeming to fit the theory into biology more generally, that was an error in so far as some cells can fuse to form a **syncytium**. For several decades, the doctrine looked as though it provided a base specifically for neuroscience and neuroscientists,[31] who treated it proudly as providing a special foundation for their subject (see reviews by Shepherd (1991), Bullock et al. (2005), and Guillery (2005, 2007). It was heavily dependent on the observations made by Ramón y Cajal, the Spanish neuroanatomist, who, as one of the last of his many important contributions to neuroscience, wrote a detailed account supporting the theory and demonstrating the supposed errors of the theory's opponents, the reticularists (Ramón y Cajal 1954). There were many supporters of each view, claiming many different examples of relationships that appeared to support neural discontinuities for the **neuronists**, and continuities

---

[31] Although they were not called neuroscientists until much later.

for the reticularists. Ramón y Cajal and the Italian histologist, C Golgi, a leading reticularist, both received the Nobel Prize in 1906, and the argument, which even formed a part of the award ceremony, continued for many years thereafter.

When I started as a student, the neuron doctrine or theory had played a fundamental role in our understanding of nervous systems for more than 60 years, and it provided a key for tracing the major functional connections of the brain. Much of the anatomy in this book was first described on the basis of the neuron doctrine, but today the doctrine plays a limited or no significant role except in historical accounts. This change is important for understanding the theoretical underpinnings of neuroscience and where it stands in relation to biology.

The neuron doctrine saw each nerve cell as having a single, readily recognizable long process, the axon, suitable for carrying messages away from the cell body over long distances, and several shorter pro-cesses that branch close to the cell body rather like the branches of a tree, the **dendrites**, for receiving inputs (see Chapter 2, Figure 2.4 and Figure 5.2(a, b)). The axon most commonly arises from the cell body, but occasionally is given off by one of the dendrites.

An important contribution to the strength of the neuron doctrine was called the **law of dynamic polarization**, hinted at earlier by Deiters (see Deiters 1865; Deiters and Guillery 2013) and fully developed in the 1880s and 1890s by Ramón y Cajal and Van Gehuchten (Berlucchi 1999). This stated that the cell body and the dendrites receive inputs from other nerve cells whereas the single axon serves to send outputs to other cells. The messages carried by the nerve cells were transmitted in one direction only from one cell to another at neural junctions, named synapses by Sherrington. Here the parts of two nerve cells appeared, according to the neuronists, to be in close contact with each other, but are never fused (Figure 5.3(a, b)). The neuron doctrine stated that there is a visible gap between the presynaptic and the postsynaptic elements of these junctions (Figure 5.3(a, b)), and that the synapse, like the nerve cell, is polarized, transmitting messages in one direction only. The reticularists argued for continuity, and generally did not recognize the polarization.

(a)   (b)

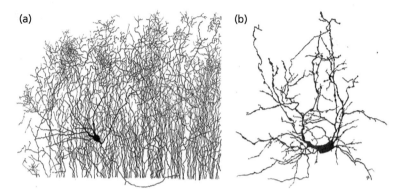

Figure 5.2 Two cells from the thalamus of a mouse.

(a) A single large cell from the lateral geniculate nucleus of a mouse, its cell body probably about 20μm in diameter, stained by the Golgi method. The cell lies among incoming axons from the retina. The cell has several dendrites that receive inputs (not shown) from the retina and has a single axon emerging at the bottom of the figure on its way to the cortex. Notice that this axon originates from a dendrite close to the cell body.

From Kölliker (1896).

(b) A smaller cell from the lateral geniculate nucleus of a mouse, about 10μm in diameter, at a higher magnification, stained by the Golgi method. This cell represents a thalamic 'interneuron', small cells whose axons ramify locally, close to the cell body but commonly lack an axon. The dendrites have an unusual axon-like appearance, and can function like an axon as presynaptic parts of a synapse (see Chapter 8). That is, the dendrites serve not only as a surface for receiving inputs but also, at local swellings provide an output, presynaptic to other local dendrites. The problem presented by axonless cells, also present in other parts of the brain, and known already to Ramón y Cajal's generation, that is, of a cell that apparently has no output, was not resolved until the 1960s when the electron microscope showed that these dendrites could be presynaptic elements of synapses.

From Kölliker (1896).

The neuronists' claim was significantly based by Ramón y Cajal on a method of staining nerve cells that had been developed by Golgi.[32] This, the **Golgi method**, stained only about 1% of the nerve cells and fibres in a tissue slice, but those cells were generally completely revealed, including axons and dendrites, against an almost clear background

---

[32] The Golgi method was discovered by Golgi, who preserved some nervous tissue in a dichromate solution, a method of preserving tissues adapted from the tanning industry,

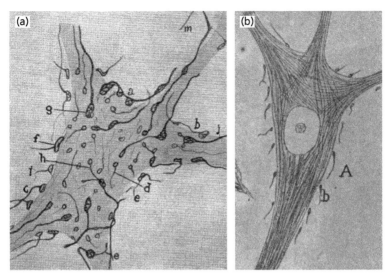

Figure 5.3 Illustrations of axons forming synaptic terminals on the surface of a nerve cell.

(a) A large nerve cell and its five main dendrites from the spinal cord of a dog. These large cells are about 30–60µm in diameter and send their axons out from the spinal cord to innervate muscles. Note the many incoming axons that contact the cell at their terminal 'synaptic boutons'. Where they are at the border of the cell's profile and don't overlap the surface of the cell (top left, bottom right, also f and i), they can be seen to be separated from the cell surface. The cell was stained by a reduced silver method which here stains primarily the fibrillary component of the axon terminals. Part (a) only shows very faint fibrils in the cell body, in contrast to the cell in part (b) whose fibrils are well stained. The vagaries of these methods were well recognized but their chemical basis was never understood; to some extent they represent the vagaries or the ways in which the fibrillary components are distributed in nerve cells (see Chapter 7).

From Ramón y Cajal (1954).

(b) A large nerve cell from the spinal cord of a lizard stained by the same method as (a). This would probably have been smaller than the cell in (a) but it was cut so that most of the contacts are seen at the borders of the cell, the surface of the cell appearing in an adjacent section.

From Ramón y Cajal (1955).

---

and then transferred the pieces to silver nitrate, a solution known for staining nerve cells, and based on early photographic procedures. The result, a relatively complete black impregnation of a small proportion of the nerve cells in the tissue, was completely unexpected and wonderfully revealing of neuronal structures.

of unstained cells (see Chapter 1, Figure 1.1(e) and Figure 5.2(a, b)). Ramón y Cajal used Golgi's method extensively in his studies, and made the important observation that the mysterious 'black substance', which impregnated the cells, never crossed from the processes of an impregnated nerve cell to that of another, un-impregnated, cell, even where other methods, which did not pick out just a small proportion of cells, showed close synaptic contacts when both cell processes had been stained (Figure 5.3). This can be regarded as one key hypothesis of the neuron doctrine: nerve cells are discontinuous, polarized units.

Some of these other methods, called **reduced silver methods**, showed a fibrillar element of the nerve cells, the **neurofibrils**, which varied significantly in their distribution but generally revealed axons and dendrites and the cell body (see Figure 5.3(a, b)). In many places, they also revealed the synapses showing presynaptic fibrillary rings or clubs, called '**synaptic boutons**' by Ramón y Cajal, and often, where the pre- and postsynaptic processes were not overlapping within the thickness of a section, they would reveal a clear 'synaptic gap'.

Details of how messages are transmitted along axons by brief changes in the electrical potential that passed along the axon (the action potential), were worked out later (Hodgkin and Huxley 1952a, 1952b), as were details concerning the transfer of the messages from one nerve cell to another (see Chapter 7).

The neuron doctrine combined with the law of dynamic polarization provided a powerful analytical tool for tracing the pathways of the brain. Ramón y Cajal used it to produce many useful figures in which some of the major pathways of the brain were represented by nerve cells linked to each other along pathways forming a clear one-way system indicated by Ramón y Cajal's characteristic arrows (Figure 5.4(a, b)). This way of understanding the pathways of the brain was tremendously powerful and provided a basis for studies of neural connections for more than 70 years.

The neuron doctrine itself limited the pathways along which messages could be expected to spread: the messages could only pass from one nerve cell to another at the synaptic junctions, which provided a one-way passage from axon to dendrite or cell body.

(a)   (b)

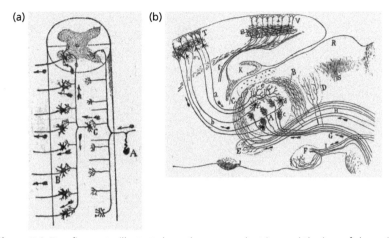

Figure 5.4 Two figures to illustrate how the neuron doctrine and the law of dynamic polarization allowed Ramón y Cajal to define the one-way conduction along axons and dendrites.

(a) Pathways entering and leaving the spinal cord. Ramón y Cajal's A (half-way down the page on right) shows the dorsal root ganglion cell with the peripheral axon on the right bringing messages from the skin, the muscles, or the deep tissues into the spinal cord. Ramón y Cajal's C (among the cells of the spinal cord) shows the spinal cells that pass messages on to higher centres.

(b) The sensory pathways from the body and the limbs traced schematically through the thalamus to the cortex. The cortex is at the top of this figure and the thalamus is in the middle (Ramón y Cajal's A). The lower parts of the figure show the ascending fibres going from the brainstem to the thalamus and giving off branches to another centre (D) on the way.

However, this view was strongly opposed by supporters of the '**reticular theory**', the **reticularists**. They considered that the processes of any one nerve cell can fuse with those of many other nerve cells, leaving no narrow gap in the contact region (see Figure 5.5)[33] and linking, directly or indirectly, each nerve cell to a vast and generally unanalysable collection of other nerve cells. This significantly lacked the sense of direction, provided originally by the law of dynamic polarization and later by the

---

[33] I have used this illustration to represent many different preparations that convinced reticularists that neuronal junctions represented cellular continuities. The arguments were complex, detailed, heated, and often fascinating. The details shown in the figure are discussed further in Chapter 7.

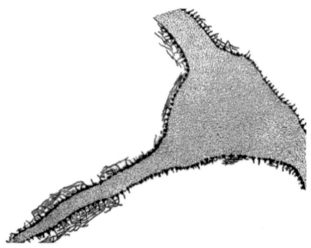

Figure 5.5 A large motor cell from the brainstem of a rabbit. Note the very much denser distribution of synaptic terminals than that revealed in Figure 5.3(a) for a cell that is functionally roughly equivalent to the cell shown here. The difference in staining methods accounts for the striking difference in the appearance of these two preparations. This is discussed further in Chapter 7.

Redrawn from Auerbach (1898).

demonstration that synapses provide a one-way path of communication between nerve cells.

Whereas the neuron doctrine provided a start for a reductionist approach to the brain, identifying the individual functionally significant parts and pathways of the brain, the reticularists were holistic, essentially arguing that the unity of the brain could not be analysed in terms of separate distinct functional parts. Golgi, the leading reticularist, wrote in his 1906 Nobel Prize speech, 'I cannot abandon the idea of a unitary action of the nervous system'. In contrast, Ramón y Cajal's schematic drawings (Figure 5.4(a, b)) showed the parts that provided the action and are still useful today to students who are no longer concerned about the neuron doctrine.[34] The argument between

---

[34] Ramón y Cajal (1937, page 338) wrote: 'To affirm that everything communicates with everything else is equivalent to declaring the absolute unsearchability of the organ of the soul'. The sarcasm of this statement is hard to miss when one compares it with Golgi's statement cited earlier in the paragraph.

the reticularists and the neuronists lasted for well over 50 years, and included some fierce and outspoken disagreements.

Shortly after JZ had published his description of the fused axons in the squid, the leading histology textbook, which strongly supported the neuronists' view, argued that if the giant fibres were processes of several different cells that had fused, then, by definition, they could not be the processes of nerve cells. JZ, also a strong supporter of the neuron doctrine, wrote: 'It is not necessary to delay over the question of whether we should save the letter of the neuron theory by saying that such cells are, by definition, not neurons (see Maximow and Bloom, 1938). It is important to recognize that the occurrence of such fusions *does not invalidate the neuron theory in general*' (Young 1939, his italics).[35]

This is JZ at his dogmatic best, cutting through years of complex apparently theoretical arguments about an important basis of the subject. He saw the neuron doctrine as a tool, not as a final truth about the nature of the world.[36] He did not treat it as a hypothesis that had been falsified by the fusions that produced the squid giant fibres. Later, Hodgkin and Huxley (1952a, 1952b) used squid giant fibres to demonstrate the details of how action potentials are generated in axons. No one then or (to my knowledge) since has raised the issue that the squid's giant axons might not be nerve fibres!

5.3.2 The importance of descriptions: neuroscience as biology

JZ's approach to the neuron doctrine was an important lesson for me. The neuron doctrine would, a few years later, be treated by many as a

---

[35] Maximow was the leading histology textbook of the time.

[36] Paul Nurse, the then president of the Royal Society, wrote recently (Nurse 2015): 'Science is a high calling in the pursuit of truth'. However, he also recognized that 'scientific knowledge evolves and may be rather tentative ... Science advances by constant testing and eliminating unsatisfactory ideas and hypotheses'.

It is important to recognize that in so far as we are in pursuit of the 'truth', we are in pursuit of the most reliable observations and their most reasonable and logical interpretations. We are not in pursuit of some ultimate, absolute, unchangeable 'truth' about what the world is 'really' like.

testable hypothesis rather than a doctrine or a theory. This was a view stressed at the time by K Popper (1959)[37]: that science is essentially about testing hypotheses and that hypotheses could not be proved by any number of supporting observations but could be destroyed by a single negative one. This, called the principle of falsifiability, was supposed to demarcate science and distinguish it from studies, such as those of Freud, which were classified as not falsifiable and therefore not science. PB Medawar who was appointed to the chair of the Zoology department at University College London in 1951, was later a strong supporter of this view (Medawar 1969). He saw the early rapid progress of molecular and cell biologists as they worked to reveal the genetic code and its role in protein synthesis as a clear example of the power of Popper's approach, as, indeed, it was.

Later, when I was working with Brian Boycott, who was then a member of the zoology department and had previously worked with JZ on the octopus brain, Medawar at the departmental tea breaks, which then formed a far more important social function in English universities than they do today, would teasingly challenge Brian about his studies of the brain, asking if there were any clear, testable hypotheses about the brain. For us at the time it was useful to know that JZ saw that we lacked clear questions or hypotheses about the brain and that descriptive science was a good way to generate questions. It was not until we had found and then answered some of the questions, and defined some of the areas of ignorance, that we might be ready to formulate specific hypotheses. Much later, when Huda Akil was president of the Society for Neuroscience, she wrote a fine review (Akil 2003) in the society's quarterly magazine in 2003 stressing the importance of what she called discovery science, and mentioning the human genome project as a good example.[38]

---

[37] Popper's view should be contrasted with the JB Conant statement cited in the second to last paragraph of this chapter.

[38] The art of expressing a putative scientific investigation as a test of specific (fundable) hypotheses has been greatly stimulated by the rules that applicants for research grants are often expected to follow. The recent account of NICHD's (National Institute of Child Health and Development) decision to design a study as a 'data collection platform', guided

Specifically, if in 1936, when JZ described the squid giant axons as the fusion of several smaller axons, the neuron doctrine had been regarded as a falsifiable hypothesis that said nerve cells do not form fusions, then, according to Popper, JZ's account of the neural fusions would have been taken to spell the death of the neuron doctrine in spite of the fact that this view of nerve cells as the unidirectional building blocks of the nervous system was still extremely useful in 1936, and would continue to be useful up to the present. Accepting the available alternative, the reticularists' view would have limited the answerable questions; if everything is connected to everything else then one cannot ask questions about neural connections, and the arrows that characterize many of Ramón y Cajal's drawings (see Figure 5.4(a, b)) could never have been used in analysing the fibre pathways of the brain.

Although the neuron doctrine had been named as though it had to be true, it was in practice not really a dogma, a theory, a hypothesis, or a belief, it was more like a set of rules about what most nerve cells are like; it was a useful piece of descriptive science that could be interpreted in terms of some important functional properties (Guillery 2005, 2007). The neuron doctrine did not describe what the world is 'really' like. It was not, as Shepherd (1991) in his account of the history of the neuron doctrine claimed, that the neuron doctrine was one of the great ideas of modern thought, comparable to the quantum theory, relativity, and the theory of evolution. Had Shepherd treated any part of the neuron doctrine as a falsifiable hypothesis, he could not have made this claim in 1991, many years after he himself had shown that dendrites could be presynaptic as well as postsynaptic (Rall et al. 1966),[39] and others had demonstrated channels of communication between nerve cells at specialized ('gap') junctions (Furshpan and Potter 1959; Bennett et al. 1963).

---

by just a few hypotheses—'such as that exposure to kitchen dust exacerbates respiratory problems' (Kaiser 2014, page 1327)—may prove hard to falsify, and illustrates the degree to which the art of formulating and testing hypotheses has degenerated relative to what was intended by Popper and Medawar.

[39] Thus falsifying the law of dynamic polarization.

In 2005, Bullock and colleagues published a paper in *Science*[40] in which they stated that: 'After a century, neuroscientistis are rethinking the Neuron Doctrine, the fundamental principle of neuroscience' (Bullock et al. 2005, page 791). They concluded that: 'The (neuron) doctrine ... no longer encompasses important aspects of neuron function', and that 'Information processing in the nervous system must operate beyond the limits of the neuron doctrine'.

The neuron doctrine had played an extremely important role in showing Ramón y Cajal and many others how one could start to analyse the circuits and pathways of the brain. It would prove useful for a few decades yet when I was starting, and it still had a useful role to play in 2005, when Bullock and colleagues published their critique. Their paper did not make the stir in 2005 that one might have expected from Shepherd's earlier statement, which itself had also raised few arguments at the time. Many neuroscientists, like many other scientists, were doing science without being overly concerned about the rules.[41] They were defining the borders of the known, looking for areas of ignorance that needed study. Looking back at the arguments at that time, it is clear that, in spite of claims to the contrary, most people were not working according to the rules defined by the theoreticians; we were not looking for falsifiable hypotheses. We were trying to understand the bits of the brain we could see under our microscopes.

The idea that good science can be 'demarcated' in terms of falsifiable hypotheses is one that has unduly influenced much science thinking, writing, editing, as well as grant reviewing. It is worth stressing, not

---

[40] They gave several reasons, but the most relevant at the time was that functional electrical coupling between nerve cells had been shown to occur at '**gap junctions**' that joined two neural processes and allowed the diffusion of small molecules from one nerve cell to another (the dye Procion˚ Yellow was used early on to demonstrate this). That is, these were the neuronal fusions denied by the neuron doctrine. There were several other reasons including the law of dynamic polarization that could not apply to dendrites that are presynaptic.

[41] The rigorous studies that had described axonal conduction of nerve **impulses** (Hodgkin and Huxley 1952a, 1952b) or of synaptic transmission (Fatt and Katz 1951, 1952) at that time were clear examples of the sort of science that Medawar and Popper had been describing, and they will have a longer life than the neuron doctrine.

only that descriptive science has an important role to play in developing scientific fields, but also that the falsification of a hypothesis only rarely destroys it; there are usually ways to counter the apparent falsification, as was strikingly illustrated by the Maximow and Bloom textbook when it simply ruled that squid giant fibres were not parts of nerve cells! JB Conant (1947, page 44)[42] had earlier, rightly, pointed out that 'A theory is only thrown over by a better theory, never merely by contradictory facts', and later Kuhn developed this further by recognizing that an area of science could often live with several hypotheses that had been 'falsi-fied'; it was not until there was a major shift in the understanding of the broad field, which Kuhn called a paradigm shift, that the old falsified hypotheses could be rejected (Kuhn 1962).[43]

So far as the main issues addressed in this book are concerned, com-paring the evidence for the standard view of sensory-to-motor process-ing with the interactive view of sensorimotor integration, we have to recognize the standard view as the accepted paradigm of the past several centuries, to recognize that this paradigm raises some puzzling issues that make a search for an alternative integrative paradigm seem worth-while. However, at present, neuroscientists are lost between the two paradigms and this book represents an effort to clear a space for a new but currently still ill-defined paradigm. I will be pointing out issues that need to be resolved, questions that need to be asked because the interac-tive view has brought them into focus. Generally, we are not ready for ambitious hypotheses; we need to clarify large areas of ignorance about neural pathways, particularly the cortical motor outputs that form a part of the hierarchy of transthalamic corticocortical pathways.

It took me a long time to work out what I had learnt about the scien-tific method during my time as a student with JZ. In retrospect, I can recognize one important thing that he taught me: not to trust any recipe for doing science. TH Huxley's statement that 'Science is nothing but

---

[42] This reference and Helmholtz's study of the visual sense provide wonderful introductions to thinking about the ways in which science is done.

[43] Max Planck (1949) had a different view: 'A new scientific truth does not triumph by con-vincing its opponents and making them see the light, but rather because its opponents eventually die, and a new generation grows up that is familiar with it'.

trained and organized common sense'[44] comes close to my present views although I prefer the statement of Percy Bridgman, 1946 Nobel Laureate, for his work in high-pressure physics: 'The Scientific Method is doing your damnedest, no holds barred!' (quoted by Steinbach 1998). Much of what follows raises questions about the structures and functions of brain parts that are important for an interactive view of the brain based on a hierarchy of neural centres dominated by the cerebral cortex. These are questions that do not arise in the standard view.

---

[44] http://www.quotationspage.com/quote/26215.html

Chapter 6

The mamillothalamic pathways: my first encounter with the thalamus

6.1 **Summary**

My thesis studies stimulated an interest in the mamillothalamic pathways but also some puzzlement because we knew nothing about the nature of the messages passing along these pathways. Several laboratories were studying the thalamic relay of sensory pathways with great success during my post-doctoral years. Each transthalamic sensory relay could be understood in terms of the appropriate sensory input, but we had no way of knowing the meaning of the mamillothalamic messages. I introduce these nuclei as an example of the problems that need to be addressed for the many thalamic nuclei about whose input functions we still know little or nothing.

Early clinical studies of mamillary lesions had suggested a role in memory formation, whereas evidence from cortical lesions suggested a role in emotional experiences. Studies of the smallest of the three nuclei forming these pathways then showed it to be concerned with sensing head direction, relevant but not sufficient for defining an animal's position in space. More recent studies based on studies of cortical activity or cortical damage have provided a plethora of suggestions, each depending on the questions asked. That simple conclusion is relevant for all transthalamic pathways. The evidence introduced in Chapter 1, that thalamocortical messages have dual meanings, suggests that we need to rethink our questions. It may prove useful to look at the motor outputs of relevant cortical areas to get clues about some appropriate questions.

6.2 **The hypothalamus and the mamillary bodies**

My studies of the hypothalamus had a difficult start. Its cells are small and the axons that connect the parts of the hypothalamus to each other and to the rest of the brain are unusually fine,[45] hard to trace, and also often hard to stain. I spent much time learning about the staining methods that could reveal nerve cells and nerve fibres, but because most of the methods had been available for the past 50–75 years, I could learn nothing new about the hypothalamus.

I decided to focus on two, well-defined structures that formed in the most posterior part of the hypothalamus, one on each side: the mamillary bodies (see Chapter 1, Figure 1.1(b, c), and Figure 6.1). They are connected to other brain regions by well-defined fibre bundles that had already been clearly described in the nineteenth century.

The mamillary bodies form a part of an interlinked circuit of connections often referred to as the **Papez circuit** (Figure 6.1). The mamillary bodies (labelled 18 in Figure 6.1) send fibres to the principal mamillary tract (labelled 15 in Figure 6.1) and the **mamillothalamic tract** (labelled 11 in Figure 6.1) passes to three anterior thalamic nuclei (labelled 8 in Figure 6.1), which are not shown separately in the figure. The anterior thalamic nuclei, in turn, send their axons (not illustrated) to three cortical areas, the **anterior cingulate cortex**, the **posterior cingulate cortex**, and the **retrosplenial cortex** (labelled 1, 2, and 3, respectively, in Figure 6.1). These cortical areas all send fibres through two parahippocampal areas to the **hippocampus**, formed by the two C-shaped structures (H). The hippocampus, in turn, sends fibres through the **postcommissural fornix** (labelled 7 in Figure 6.1) to the mamillary bodies.

The pathway from the anterior thalamic nuclei to the cortex[46] resembles the sensory relays of the thalamus that relay visual, auditory, and

---

[45] Thin nerve fibres conduct impulses slowly, thick fibres conduct more rapidly. The visceral nervous system generally does not need rapid reactions.

[46] There are three separate nuclei, anterior medial, anterior ventral, and anterior dorsal, not individually shown in the figure, which connect to the cortical areas labelled 1, 2, and 3 (anterior cingulate, posterior cingulate, and retrosplenial), respectively.

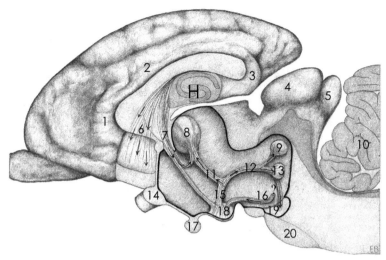

Figure 6.1 A view of some of the major parts of the mamillothalamic and mamillotegmental connections (often called the 'Papez circuit') in the rabbit, based on a figure from Kölliker (1896). The figure shows the medial aspect of a rabbit's cerebral hemisphere with the pathways that link the mamillary bodies to the midbrain and the anterior thalamic nuclei exposed by removal of surrounding tissues adjacent to the third ventricle. The nuclei and the pathways connected to them have been dissected to show the major connections of the mamillothalamic circuit. Note that the subiculum and the entorhinal cortex which link the retrosplenial cortex (3) to the hippocampus (H) are not shown in this figure, and the mesencephalic origins of the mamillary peduncle (shown as cut and depicted with a '?' to indicate that all inputs are not yet known) are not illustrated; only those from the dorsal and deep tegmental nuclei are known, but there may well be others (see text).

1, anterior cingulate cortex; 2, posterior cingulate cortex; 3, retrosplenial cortex; 4, superior colliculus; 5, inferior colliculus; 6, precommissural fornix; 7, postcommissural fornix; 8, anterior thalamic nuclei; 9, dorsal tegmental nucleus; 10, cerebellum; 11, mamillothalamic tract; 12, mamillotegmental tract; 13, deep tegmental nucleus; 14, optic chiasm; 15, principal mamillary tract; 16, mamillary peduncle; 17, pituitary gland; 18, mamillary body; 19, tegmental reticular nucleus of the pons; 20, pontocerebellar fibres.

Illustration by Dr Lizzie Burns.

other sensory messages (other than olfactory) from the thalamus to the cortex, although for the two major anterior thalamic nuclei we still need to define the nature of the messages that the anterior thalamus relays to cortex. The smallest of these nuclei, the anterodorsal, sends information to the cortex about the position of the head.

The mamillothalamic pathways will here serve to illustrate some of the fundamental problems that need to be resolved for a thalamic nucleus where specific information about the nature of the inputs is lacking; that is, for the majority of the thalamic cells in a primate's brain.

6.3 The functions of the mamillary circuit

I had to think about what the mamillary pathways might be doing. We had been taught that at post-mortem, brains of individuals who had died at late stages of chronic alcoholism show degenerative changes in the region of the mamillary bodies or their output through the principal mamillary tract (Figure 6.2) and its main mamillothalamic branch (Figure 6.1). **Korsakoff** (1887) had related these changes to memory losses, which accompany this condition. No one knew anything about the specific role that the pathways played in the formation, maintenance, or recall of memories; that was evidence produced some years later by Scoville and Milner (1957; and see Squire 1992), who showed that the hippocampal formation, which sends many of its outputs to the mamillary bodies in the postcommissural fornix (labelled 7 in Figure 6.1), plays an important role in the formation of new memories.

6.3.1 The role of the mamillothalamic relay to the cortex

I have already indicated that one fundamental issue for understanding any thalamic relay concerns the questions we ask about the functions of that relay. Those questions will define, and will also limit, the answers we get. This is clearly demonstrated by our knowledge about the anterior thalamic relays of the mamillary inputs, where the functions of the relay cells in the two largest anterior thalamic nuclei, containing the vast majority of the anterior thalamic relay cells, are still undefined or defined by a set of arbitrarily chosen questions that have produced a strange hodgepodge of answers (see Table 6.1 later in this chapter). These answers, as we will find, have provided some clues about the functions of these two largest anterior thalamic nuclei, but nothing that appears to be as clear as the evidence for the smallest of the anterior thalamic nuclei, the anterodorsal nucleus, where in the 1990s a function

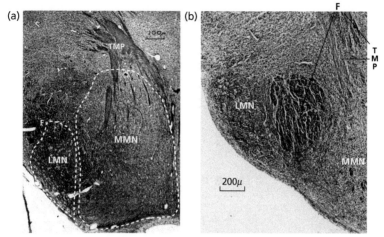

Figure 6.2 Two transverse sections through the mamillary bodies of a cat (a) and a rabbit (b) showing the large medial (MMN) and the smaller lateral mamillary nucleus (LMN). The reduced silver method used to stain these sections shows the nerve fibres that form the major fibre bundles (the fornix, F; and the principal mamillary tract, TMP) and at this magnification shows some of the pale relatively large nerve cells in the LMN.

Section (a) passes near the back of the mamillary bodies, where most of the fibres of the fornix have passed to their terminals within the mamillary bodies. Some of these fibres are curving into the lateral parts of the medial mamillary nucleus and others form the bundle labelled F and are destined for the posterior parts of the mamillary nuclei.

Section (b) lies at a level anterior (ahead of) section (a), where the fornix forms a well-defined bundle and the fibres of the principal mamillary tract (TMP are gathering to form its main ascending bundle. Notice the fornix lying between the two nuclei in the rabbit, but not in the cat.

Reproduced from R.W. Guillery, 'The mamillary bodies and their connections', PhD thesis, University College London, 1954.

was defined and traced back to its sensory origins in the vestibular organs, whose cells are activated by movements of the head. These **'head direction cells'** are discussed in Section 6.4.

6.3.2 **The likely functions of the mamillary circuit**

I had to think about what the mamillary pathways might be doing. Apart from Korsakoff's suggestion that the mamillary bodies play a role in the formation of memories, another suggestion had been made by Papez (1937) on the basis of the functional losses that accompany lesions of the **cingulate cortex** (labelled 1, 2, and 3 in Figure 6.1). Papez proposed that the mamillary bodies might be concerned with the conscious awareness of one's own emotions: that cells in the mamillary bodies might be receiving copies of messages that were known to pass from the hypothalamus just ahead of the mamillary bodies (labelled 11 in Figure 1.1(b)), downstream towards the spinal cord for the control of sweating, blood pressure, heart rate, gut contractions, bladder contractions, respiratory rate, lachrymal secretions, and so on (Loewy and Spyer 1990). These are all called 'visceral' actions and are controlled through special parts of the **peripheral nervous system** called the **sympathetic** and **parasympathetic systems** (or more generally the **visceral nervous system**). One or more of these actions relates to particular emotional experiences,[47] and copies of these instructions for visceral outputs would then, according to Papez's suggestion, be passed by branches of the hypothalamic output fibres to the mamillary bodies and through the anterior thalamic nuclei to the cingulate cortex, bringing information about the emotional experiences to the cerebral cortex and thus creating an awareness of the emotional experiences.

This suggestion, that copies of hypothalamic motor outputs that are sent to the cortex can provide a mammal, through its mamillothalamic pathways, with a sense of its own *forthcoming* (visceral) actions, can today be seen as a precursor of the view introduced in Chapter 1 that the messages that the thalamus relays to the cortex generally include information about forthcoming actions and their sensory consequences. Since the nerve fibres that pass messages to the viscera conduct messages slowly, and the visceral organs generally do not respond rapidly, while the mamillothalamic pathways conduct more rapidly and involve a shorter distance, awareness of the forthcoming emotional response is likely to be significantly ahead of the visceral action itself.

---

[47] A theory of the emotions called the James–Lange theory, had earlier proposed that the visceral response was primary and the emotion was produced by the experience of that response. The evidence for the necessary hypothalamic connections to the mamillary bodies is currently lacking, but should not be difficult to obtain.

These two apparently quite different suggestions for a possible role of the mamillary bodies, awareness of the emotions as opposed to memory formation, remain today as a fascinating challenge to contemporary neuroscience. A possible start would be to search for the required functional connections from the hypothalamus to the mamillary bodies in relation to a record of the activity of the relevant cells in various emotional situations. The methods for studying these connections, functionally and structurally, had not been developed in the early 1950s, but are now available.

When I started on my thesis research, brains were starting to be compared to computers. Quantitative relationships between the cells and their connecting axons in linked neural centres became of interest. The numbers gave such studies an air of contemporary science: it was thought that they might provide insights concerning the connected cell groups. Others had studied some of the quantitative relationships of the nerve cells on the pathways that link the photoreceptors in the retina through several other retinal cells to the thalamic relay for vision, the lateral geniculate nucleus (Walls 1953). It seemed possible that the rather well-circumscribed structures of the mamillary bodies and of the pathways linking them to the rest of the brain (Figure 6.2) might also reveal some important quantitative relationships between the parts of this circuit. I produced a detailed review of the literature on the mamillary bodies and their connections, and thanks to my knowledge of German was able to summarize a large part of the older German literature about the hypothalamus. That not only helped to give my thesis an appearance of solidity that the original science lacked, but also established a lifelong interest in the history of the subject as well as a love of browsing through old books on open library shelves that were then welcoming even to a mere student. These 2 years also introduced me to the puzzles of thalamic functions. What is the thalamus for? Why don't messages pass straight to the cortex from the ascending sensory pathways?

The counts and a serendipitous discovery, that one of the pathways feeding into the mamillary bodies (the postcommissural fornix, see Figure 6.1 (label 7) and Figure 6.2) lost many of its axons on the way to the mamillary bodies (Guillery 1956), provided me with a thesis, left me with an interest in the mamillary bodies, and was an excellent subject for my first postdoctoral research: to find out what happened to the fibres that did not reach the mamillary bodies. It turned out that they pass to the thalamus and there join the direct projection from the mamillary bodies to the anterior thalamus (see the pathway turning from 7 to 8 in Figure 6.1). This provided a start for my career but the main problem, of what messages the anterior thalamic nuclei were sending to the cortex, remained unsolved, as did the more general problem concerning the function of the thalamus. We could ask the questions, but needed to find a specific way to approach the problem experimentally.

6.4 Later advances: the head direction cells and the anterodorsal thalamic nucleus

Our view of the mamillothalamic pathways was singularly altered during the 1990s and later, by a remarkable series of publications by J Taube and co-workers (Taube 1995, 2007; Dumont and Taube 2015). These focused mainly on the smallest of the three anterior thalamic nuclei, the anterodorsal nucleus, and showed that in rats these cells respond to the direction in which the head is pointing. They were called 'head direction cells'.

Consider a rat that is exploring its environment, perhaps an experimental maze, but more realistically, though less often studied, an inner city waste lot. It will stop and turn its head one way or the other as though making a decision about its best next move, perhaps recording the scene in its memory.

These head direction cells were also shown in the retrosplenial cortex (labelled 3 in Figure 6.1), which receives inputs from the anterodorsal nucleus and passes messages through two other cortical areas, **subiculum** and **entorhinal cortex** (not shown in Figure 6.1) to 'place cells' of the hippocampus (labelled H in Figure 6.1). These fire maximally only when the rat is in one particular part of the environment (O'Keefe and Dostrovsky 1971; O'Keefe 1978). In addition, and most importantly for understanding the biological origins of the message, Taube and colleagues asked where the message originated and traced it back from the anterodorsal nucleus towards its origins: first to the lateral mamillary nucleus (see Figure 6.2) which had long been known to provide **afferent** axons to the anterodorsal thalamic nucleus (e.g. Kölliker 1896; Cajal 1955), and then further back to the **dorsal tegmental nucleus** in the midbrain (labelled 9 in Figure 6.1),[48] a nucleus that had also been described already in the nineteenth century as sending inputs to the mamillary bodies. In the dorsal tegmental nucleus, Taube and colleagues recorded cells that integrated messages from the **vestibular nuclei**. These lie more caudally in the brainstem, and themselves play a major role in controlling head movements.

These studies of head direction cells represent a wonderful persistence in tracking a novel function to its sources through an interconnected series of neural centres, demonstrating the messages that are being passed sequentially through several different cell groups from the vestibular nuclei and the dorsal tegmental nucleus in the brainstem, to the mamillary part of the hypothalamus, the thalamus, the retrosplenial cortex, and eventually to the hippocampus. The message about head direction is generated in the dorsal tegmental nucleus, on the basis largely or entirely of the vestibular inputs, and is then transmitted through the mamillary bodies and the anterior thalamus to the retrosplenial cortex. The general resemblance to other thalamic nuclei, which receive inputs about visual, auditory, somatosensory, or gustatory inputs is clear: the striking feature of all of these relays is that, whereas most of the relays on the way to the cortex are modifying the message, the thalamic relay (and here the relay in the lateral mamillary nucleus as well) seems to be doing nothing. This, as indicated earlier, would be one of the challenges for those of us interested in the thalamus. What is the thalamus for?[49]

---

[48] This pathway is shown as cut in the figure and indicated by a '?'; it carries a mixture of inputs to the mamillary bodies in addition to that from the dorsal tegmental nucleus.

[49] In the late 1960s or early 1970s, there was an exhibition in the Natural History Museum in South Kensington, London, showing the functional organization of the visual pathways from the retina to the cortex. Oddly, there was no thalamic relay in this schema, and at the time this omission made no difference to the way the visual pathways were understood. We laughed about this omission at the time, but we had no answer to the question: 'What is the thalamus for?' The problem for the thalamus is addressed in the remaining chapters of this book, but the lateral mamillary nucleus remains as a puzzle.

It may well prove that the postcommissural fornix serves to modulate transmission through this nucleus on the basis of ongoing hippocampal activities.

Other information may be added as the message passes from the lateral mamillary nucleus towards the hippocampus. Possibly anticipation of visceral activity is added in the lateral mamillary nucleus, as suggested by Papez. In the anterodorsal nucleus, information about visual signals in the rat's environment is added (Figure 6.3). The cells in this nucleus change their response when prominent visual features are switched from one side to the other of the rat's usual environment (Taube 2007). These added visual signals in the anterodorsal nucleus are then passed from the thalamus to cortex. At present it is not clear whether these visual inputs come from the midbrain as does the information about the head direction cells or from the visual cortex, although there is some evidence to suggest that they are coming from the midbrain since we know that the mamillary inputs are the drivers (see Chapter 2) in the anterodorsal nucleus and the cortical cells are modulators (see Chapter 2; and Somogyi et al. 1978; Petrof and Sherman 2009).

These results show that there are important transthalamic pathways for messages that travel from the sensory periphery through the midbrain and mamillothalamic pathways towards the cortex. They also show that the information in these pathways may be changing as it moves from the sensory periphery to the cortical regions, adding new items on its way: possibly adding the anticipation of the visceral actions in the lateral mamillary nucleus and adding the visual signals in the anterodorsal nucleus. Each head direction cell may be linked to a particular group of visual features in the environment that are needed to create the place cells in the hippocampus.

The rat, as it explores its environment, stopping at choice points and turning its head to assess the best next move, would have the combination of head direction, visual information, and emotional colouring to compare with earlier memories and to store for future use. These would all be available for determining the next move. However, for the lateral mamillary nuclei, this is at present merely a speculation about

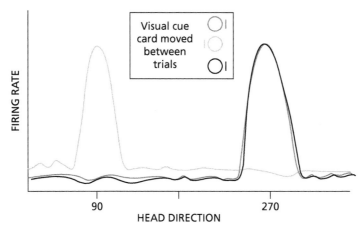

Figure 6.3 Illustration of experiments showing the role of visual stimuli in the generation of head direction cells in the anterodorsal thalamic nucleus. The record shows the firing rate of the head direction cells when a prominent visual screen is moved from the right to the left and back.

Adapted with permission of *Annual Review of Neuroscience*, 30 (1), The Head Direction Signal: Origins and Sensory-Motor Integration, Jeffrey S. Taube, p. 184, Figure 1c, DOI: 10.1146/annurev. neuro.29.051605.112854, © 2007 by Annual Reviews, http://www.annualreviews.org.

the emotional colouring being available to the rat as an *anticipation* as it turns its head at a choice point.[50]

6.5 The anteromedial and anteroventral thalamic nuclei

The anteromedial thalamic nucleus, projecting to the anterior cingulate cortex, and the anteroventral thalamic nucleus, projecting to the posterior cingulate cortex (labelled 1 and 2 in Figure 6.1), are known in far less detail than is the tiny anterodorsal nucleus, and both are

---

[50] There is one further issue that will come into focus later in this book. It concerns the nature of the activity recorded in the head direction cells. The information coming from the vestibular nuclei is not strictly speaking sensory information about the world. Should we treat it as sensory information about the body? Or is it better regarded as information about an instruction for a movement? The observation that discharges in head direction cells peak several tens of milliseconds before the optimum head direction is reached (Taube 2007; Dumont and Taube 2015) suggests that this signal may represent a copy of a vestibular motor output, as does the evidence (Ramon y Cajal 1955) that (some, many, all?) ascending fibres from the vestibular nucleus come from descending branches.

still candidates for a role in memory formation (Aggleton et al. 2010; and see also Dillingham et al. 2015) and possibly also for the cognitive processing of the emotions as suggested by Papez. The memory hypothesis as formulated by Aggleton et al. (2010) suggests that messages arrive from the brainstem for the anterodorsal thalamic nucleus but from the hippocampus for the other two nuclei. This not only looks improbable on developmental and evolutionary grounds, but is also contrary to current knowledge about drivers and modulators in the anterior thalamic nuclei (see Somogyi et al. 1978; Petrof and Sherman 2009).

Recent experiments by Vann and colleagues (Vann 2013; Dillingham et al. 2015), in which she tested the capacity of rats to learn a maze, showed that lesions of the postcommissural fornix left maze learning intact whereas lesions of the mamillothalamic tract or lesions of the **deep tegmental nucleus** (labelled 13 in Figure 6.1), which sends axons to the mamillary bodies, interfere with the maze learning. That is, the memory loss produced by mamillary lesions is not due to a loss of hippocampal inputs to the mamillary bodies but is, instead, due to injuries that prevent the information from the world and the body from reaching the hippocampus through midbrain relays. This is in accord with the observation that the major part of the loss produced by hippocampal lesions is in the formation of new memories rather than the retention of old ones (Scoville and Milner 1957; Milner 1958; Squire 1992). The hippocampus is a temporary storage of memories. Vann and colleagues' evidence suggests that all three anterior thalamic nuclei probably receive their messages about events from the midbrain for storage in the hippocampus. However, defining how the functions of each of these three thalamic nuclei differ from each other, as well as showing how they relate to each other in view of their similar connections and developmental history remains unexplored.[51]

---

[51] Two recent studies have described cells that respond to head position in response to gravitational directions in monkeys in the anterior thalamic nuclei (Laurens et al. 2016) and in bats in the retrosplenial cortex (Finklestein et al. 2015).

6.5.1 Methods for defining the functions of cortical areas and their inputs

The method just summarized for studying the anterodorsal thalamic nucleus and its projection to the retrosplenial cortex is one of the first to follow messages back towards their peripheral origins and then looking at how the messages may be changed in the forward direction, as they combine to form a complex message for action or memory storage. It differs markedly from the methods that have defined the major sensory pathways, where knowledge about the incoming sensory modality provided the initial approaches.

Another method includes studies of specific cortical areas, looking at the behavioural or cognitive effects of cortical lesions or at congenital abnormalities and relating these to increases of cortical activity (measured in terms of blood oxygen level-dependent (**BOLD**) signals, or to local electrical field potentials[52]), recorded during well-defined behavioural or cognitive acts.

There are a great many published reports of such observations for the cortical regions that receive afferents from the anteromedial and anteroventral nuclei, and Table 6.1 shows a few of the results that have been published for the anterior cingulate cortex, which receives its thalamic inputs from the anteromedial thalamic nucleus. The results in this table depend on lesion studies, on studies of field potentials, or on studies of the BOLD signal (fMRI) of ongoing brain activity.

In Table 6.1, it is the *localization* of a particular function in a specific cortical area that represents an answer to a particular question posed by the investigator.

This method clearly has significant appeal, as demonstrated by the great many publications over the past few years that have used the method. It produces reports that a particular cortical area plays a role in a named function, without concern as to how that function becomes

---

[52] Functional magnetic resonance imaging (fMRI) serves to signal increased activity in a cortical area, based on records of blood flow which has been shown to relate to levels of overall neuronal activity. Studies of field potentials relate to the overall neural activity in a cortical area.

Table 6.1 Possible functions of the anterior cingulate cortex

Influencing the contractile functions of the bladder (Talan 1966)	(V)
Regulating respiration (Pogrebkova 1970)	(V)
Production of a natural smile, as opposed to a smile posed for a photographer (Damasio 2006)	(V E)
Psychosis (Fornito et al. 2009)	(E M)
Autism (Simms et al. 2009)	(E M)
Borderline personality disorder (Whittle et al. 2009)	(E M)
Spatial working memory in rats (Mendez-Lopez et al. 2009)	(M)
Geriatric depression (Gunning et al. 2009)	(E M)
Emotional processing in general (Etkin et al. 2011)	(E)
Predicting aversive events and terminating fear (Hayes and Northoff 2012; Steenland et al. 2012)	(E M)
Evaluating the cost of foraging (Kolling et al. 2012)	(E M)
The selection and maintenance of learned options (Holroyd and Yeung 2012)	(M)
Predicting the direction of a football (Wright et al. 2013)	(E M)
Instigating adaptive switches in choice (Economides et al. 2014)	(E M)
Signalling the net value of others' rewards (Apps and Ramnani 2014)	(E)

Note: each of the items in this table has been categorized as relating either to visceral events (V), to emotional events (E), or to memory (M).

established in that area. The method is based on empirical evidence but in other ways it resembles phrenology, popular in the nineteenth century, which was also concerned to localize specific functions to particular cortical areas. The modern studies resemble 'evidence-based phrenology'. They are clearly a very significant advance on the original phrenology, but they do not tell us how particular functions are generated in any one cortical area or relate to other areas and pathways, in the way that studies of the anterodorsal nucleus and retrosplenial cortex showed us some of the workings of the brain. For a neuroscientist, they fail to tell us what the relevant messages are, leave the pathways along which the messages travel undefined, do not show what individual nerve cells along the pathways are doing to the message, and depend almost entirely on the question that is asked. In short, they do not reveal anything about the neural mechanisms by means of which the brain is interacting with the world.

The items listed in Table 6.1 as V, M, and E, respectively, refer to earlier parts of this chapter, concerning visceral actions (V), the formation of new memories (M), and emotional experiences (E), and relate to the early suggestions made by Korsakoff, Papez, and Scoville and Milner. All three emerge as relevant to the studies listed in the table. However, it is hard to be more specific about the neural actions that the three classifications may indicate.

The table suggests a way in which a list of proposed functions for any one cortical area can perhaps be generated by a meta-study, summarizing many very different proposed 'questions' about the possible functions of the one area, and then these, when added together, can perhaps begin to give a rough view of the type of functions or the range of functions one could seek to understand in neural terms.

The amount of information available from a single experiment addressing a single arbitrarily chosen question is very limited in terms of neural functions. The BOLD signal depends on slow signals that last for seconds whereas the electrical signals of individual cells change over a few milliseconds. We do not know which layers of the cortical area are active, nor do we know how the BOLD signal or the electrical activity relates to the specific classes of cortical neurons: whether they are predominantly inhibitory neurons or excitatory neurons, or whether the activity mainly reflects the inputs, the outputs, or the local neurons (Leopold and Logothetis 1996; Logothetis 2002; Bartels et al. 2008; Goense et al. 2012). The functional contributions of an active group of neurons, in terms of turning a particular input into a particular output, which would be one very good way of defining the functions of a cortical area, are left uninvestigated. That is, there is nothing comparable to integration of vestibular inputs in the dorsal tegmental nucleus, the addition of a visual to a vestibular component in the anterodorsal nucleus, or the conversion of head direction cells to place cells in the hippocampus. The evidence of the neural activity tells us nothing specific about the messages being processed.

We can now suggest that the mamillothalamic tract as a whole probably carries inputs from the midbrain to the thalamus for relay to the cingulate cortex and from there to the hippocampus. It remains to define what those inputs are for the anteromedial and anteroventral nuclei, and how they relate to the head direction cells in the smallest of the three nuclei. The contributions each nucleus makes to memory formation need to be defined as do the motor outputs of the cingulate cortex.

The parts of the midbrain from which messages pass to the mamillary bodies lie close to a region mentioned in Chapter 1, called the mesencephalic locomotor centre (Sherman et al. 2015), which as the name implies, controls a significant part of an animal's movements through the environment, functionally not unrelated to control of head direction. Further, the relevant midbrain regions lie just below and close to the superior colliculus, a structure that receives a relatively complete picture of the world in relation to the body. That is, the superior colliculus is known to receive visual, somatosensory, and auditory inputs and these are topographically mapped not only in relation to the body and the world but also to each other in terms of major directions (see Chapter 4). Any small column of cells in the superior colliculus passing from the surface to the deep layers represents particular directions in visual and auditory space and also, in terms of the general localization of body parts. The superior colliculus in our vertebrate ancestors largely served as a major link between the world and the organism's behavioural outputs, and it would be reasonable to look at the messages transmitted from the midbrain to the hippocampal formation through the anteromedial and anteroventral thalamic nuclei as providing much of the information about the life of the organism that needs to be stored in memory.

An alternative to the mesencephalic pathway for messages about the world to reach the entorhinal cortex, and thus the hippocampus, is represented by cortical pathways that carry messages from the primary sensory cortical areas through several stages to the entorhinal cortex. The demonstration of ascending messages from the midbrain does not rule out this alternative or additional route. Such a route was proposed by Jones and Powell (1970) as several separate multistage connections, each originating in a separate primary sensory cortical area. This proposal was based on anatomical evidence about the pathways. Confirmation that these connections are made by driver pathways that carry the relevant sensory information to the entorhinal cortex is still needed.[53] We need to learn about the messages that are relayed to the cortex by the anteromedial and the anteroventral thalamic nuclei and we need to learn how they relate to information that the anterior and posterior cingulate areas receive from other cortical sources.

These are clearly outstanding areas of ignorance that merit studies comparable to the strikingly successful studies of the pathway through the anterodorsal nucleus. Given the anterodorsal nucleus as a model, it is reasonable to think that the mamillothalamic tract may carry a great deal of the information that the hippocampus needs in order to establish memories about a mammal's movements in the world and about its interactions with the world.[54]

6.6 The mamillotegmental tract

There are other features of the mamillary pathways that have not been considered so far. The **mamillotegmental tract** looks like the most direct mamillary output towards motor centres. It runs in an elegant, continuous course (Figure 6.4(a)) to several nuclei in the midbrain, including the dorsal and deep tegmental nuclei (Kölliker 1896; Ramón y Cajal 1955) as well as the **tegmental reticular nucleus of the pons** (Figure 6.4(b)) (Guillery 1957; Cruce 1977; Torigoe et al. 1986).[55] At

---

[53] The Professor of Physiology at University College London when I was starting my research career, GL Brown, used to comment about anatomical studies of neural connections (for which he had not much respect): 'Give me three synapses, and I can connect any one part of the brain to another.' He was probably right.

[54] One other feature about the mamillothalamic pathways that has not been mentioned is the fact that the cells of the anterodorsal nucleus are larger than those in the other two nuclei, and have thicker axons. That is, the functions of the anteromedial and anteroventral nuclei are less demanding about the speed of communication than are the vestibular connections.

[55] The evidence from the two more recent studies shows that some of these axons come from the lateral mamillary nucleus and that others come from the medial mamillary nucleus.

the time I saw this pathway in some of my sections, no one knew what the function of this pontine nucleus was. Its role in the control of gaze (Hess et al. 1989), that is, of the movements of the eyes and the head, was demonstrated later.

The mamillothalamic tract forms as collateral branches arising from the mamillotegmental tract (Figure 6.4).[56] The involvement of this motor output of the mamillothalamic system is raised here for three reasons: (1) to stress that the motor branches of thalamic driver inputs can often provide hints about the functions of incoming fibres themselves; (2) to stress the importance of the visual system in these mamillary connections; and (3) to draw attention to the fact, explored more fully in Chapters 9 and 10, that the mamillothalamic pathways resemble the many other transthalamic pathways to the cortex in having branches that innervate motor centres.

It is worth looking more closely at the functions of the pathways that link vestibular nuclei first to the dorsal tegmental nucleus and then through the mamillary bodies to a cell group concerned with the control of gaze (head and eye direction). The mamillary projections to the thalamus and cortex are a mammalian specialization; reptiles and frogs need mechanisms for gaze control, and almost certainly have them, probably in the midbrain. If we could learn more about the reptilian mechanisms concerned with the vestibular contributions to gaze control, then it may well prove possible to see the mamillothalamic pathway to the cortex as a mammalian addition that keeps the cortex informed about the vestibular control of gaze. The mamillothalamic pathway would be seen not as a sensory system about vestibular activity, nor as a sensory system about head position, but as a higher-level monitor of a phylogenetically old gaze control system. This is a view that can suggest a novel search concerning the functions of relevant cortical areas as monitors of gaze control. As indicated, our vertebrate ancestors must have had comparable subcortical mechanisms for gaze

---

[56] Suggesting that the mamillothalamic axons formed as branches of the mamillotegmental branches, which would have appeared earlier in developmental and probably in evolution.

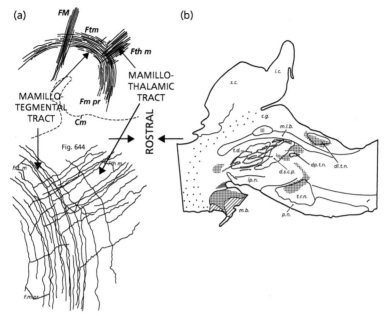

Figure 6.4 Two illustrations of the mamillotegmental tract.

(a) The fibres of the mamillothalamic and the mamillotegmental tracts as they leave the principal mamillary tract just above the mamillary bodies (not shown). The mamillotegmental axons run in a smooth curve into the midbrain, giving off the mamillothalamic branches on their way.

From Kölliker (1896).

(b) The degeneration of the mamillotegmental tract seen in the midbrain after a lesion of the mamillary bodies. Notice particularly the component passing to the tegmental reticular nucleus (t.r.n). Other abbreviations relevant to the text: m.b., mamillary bodies; dl.t.n., dorsal tegmental nucleus; dp.t.n., deep tegmental nucleus; i.c. inferior colliculus; s.c. superior colliculus.

Reproduced from R.W. Guillery, Degeneration in the hypothalamic connexions of the albino rat, *Journal of Anatomy*, 91 (1), p. 95, Figure 2 © 1957, Anatomical Society.

control and the basic puzzle of how the mammalian mamillary system developed from the hypothalamic pathways of our ancestors was clear already in the 1950s (e.g. Herrick 1948).

Today, we can also ask about the functions of the motor outputs of the relevant cortical areas. For example, for the anterior cingulate cortex the list of subcortical centres that receive an input from this area includes the following: superior colliculus, pretectum, midbrain **tegmentum**, **zona incerta**, substantia nigra, Forel field H1, amygdala, striatum, and pontine nuclei.[57] The first two relate clearly to vision. They control eye movements, and adjust the focus of the lens and the size of the pupil. The others all play a role in the control of movements, or relate to emotional reactions. However, the precise functional role that any one of these corticofugal pathways plays in relation to specific movements of the animal is another significant area of ignorance where new findings could play a key role in providing another approach to understanding the functions of a cortical area. The same point is relevant for the retrosplenial cortex and the posterior cingulate cortex. What motor actions can they influence? Possibly, knowing more about the functions of the dorsal and the deep tegmental nuclei (labelled 9 and 13 in Figure 6.1) could provide further clues about what to look for in the other parts of the mamillothalamic pathways. These nuclei send inputs through the medial mamillary nucleus to the anteromedial and anteroventral nuclei and also receive afferents from the mamillotegmental tract.

6.7 What information can one seek in order to understand the functional role of a cortical area?

1. Define the origin of the driver inputs carrying messages to the thalamus for relay to the cortical area.

2. Define the nature of the message that the cortical area receives from the thalamus, not only in terms of the past events occurring in the body or the world but also in terms of the message about an upcoming action being carried by the non-thalamic branch for motor action.

3. Define the terminations of the layer 5 outputs of the specific cortical area, and gain information about the functions of these terminal zones. This may need to be more detailed than the information currently available about these functions. Knowing that cortical area

---

[57] These are mainly parts of the motor apparatus, several of which play no further role in this book. See Domesick (1969), Sotnichenko (1976), Mueller-Preuss and Juergens (1976), and Harting et al. (1992).

A and B both project to the striatum, pons, or superior colliculus is not enough; we need to ask how A differs from B in terms not only of terminal details but also in their functional roles. Finally, where possible we need to distinguish corticofugal axons that have thalamic branches from those that lack them.

4. Look for information that summarizes lesion studies and records of activity levels in the cortical area as tested by many different approaches; that is, undertake meta-studies that can provide answers to many different questions as in Table 6.1.

If we now focus on these four points for the anteromedial thalamic nucleus we can add that the results obtained for the anterodorsal nucleus may well be relevant: the anteromedial nucleus is likely to be receiving inputs from the midbrain about the animal's movements in relation to its current environment. It may prove useful to distinguish sensory messages that follow a sensory event from instructions for forthcoming movements. These messages may have an added visual component, may perhaps also be related to copies of hypothalamic outputs, possibly providing an emotional 'colouring' to the messages that reach cortex, thus adding to the primary message, which is likely to be about the animal's movements in the current environment. Clearly, these are guesses, based on the current evidence: we need much more!

Chapter 7

Comparative anatomical studies of the hypothalamus that led to studies of thalamic synapses

7.1 **Summary**

This chapter provides a brief illustration of the role of serendipity in science. We started a comparative project to understand the differences between reptiles and mammals in terms of their hypothalamic pathways; these studies of reptilian brains revealed tiny fibrillar rings present in axon terminals of lizards kept at low temperatures but absent in those kept warmer. These temperature-dependent fibrillar increases resembled changes seen in some synaptic terminals after their axons have been cut, so we briefly turned our attention to the changing appearances of the fibrils. Mammalian **optic nerve** fibres that had been cut were known to show particularly dramatic increases of fibrillar structures. Our study of these did not show anything relevant for understanding the functions of the fibrils or their changes; instead they revealed the surprisingly complex system of synaptic structures in the thalamus, clearly demonstrating that the thalamic relay is not as simple as was generally believed. This provided a key to what follows in the rest of this book. This chapter serves to introduce synaptic structures in general, and those of the thalamus in particular. We still do not understand the differences between reptiles and mammals in terms of their hypothalamic pathways, nor do we understand the nature of the fibrillar changes.

7.2 **Reptilian brains**

Part of my interest in the mamillary bodies related to their comparative anatomy. In 1910, G Elliott-Smith, who later became the Professor

of Anatomy at University College London, published a series of lectures on the comparative anatomy of the brain (Elliott-Smith 1910).
These figured significantly in our undergraduate anatomy course and
stimulated my interest in comparing vertebrate brains, particularly of
reptiles, and mammals.

Our vertebrate ancestors had a hypothalamus but lacked the mamillary bodies, just as they lacked a thalamus linked to a six-layered cortex.
I joined Brian Boycott, who had earlier worked with JZ training octo-
pusus, in a project to study what mammals can do with their mamillary
bodies that reptiles, amphibians, and fish cannot do. Brian would do the
behavioural studies, working on the basis of memory storage and the
emotions discussed in Chapter 6, and I would do the anatomy. It was
a wonderfully productive project, but only because by sheer good luck
it led to results totally unexpected and unrelated to our original aims.

The reptiles we studied included lizards (*Lacerta viridis*); rat and cat
represented the mammals. We were told the lizards liked to be kept
warm at 32°C, but at that temperature they were so lively that we could
barely even catch them, never mind train them. So we moved some
to 19°C and found them much easier to handle. I used several vari-
ants of the reduced silver methods (see Chapter 5) to stain the lizard
brains. The results I obtained were entirely unexpected and took me on
a long route first to the study of synapses and then from the synapses
directly to the thalamus; it was like a wild goose chase ending up as a
wedding feast.

In the brains of lizards kept at 19°C, a number of small fibrous rings
were stained in several parts of the brainstem and the hippocampus (see
Figure 7.1). These rings resembled the synaptic boutons described earlier by Ramón y Cajal (see Chapter 5, Figure 5.4), but were not clearly
related to any visible stained postsynaptic structures. More puzzlingly,
they were absent in the brains of the lizards kept at 32°C.

I learnt that Ramón y Cajal had much earlier described a striking
difference between lizards caught in the Madrid winter and lizards
kept warm in the lab. His figures (Figure 7.2) show the neurofibrils in
the cell body and the dendrites. The functional significance of such
temperature-dependent changes in nerve cells, including the synaptic

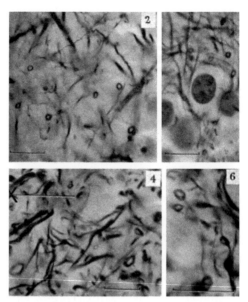

Figure 7.1 The fibrillary rings that were stained in the cold lizards (see text for details). The upper two figures are from the region of the hippocampus and the lower from the brainstem.

Reproduced from B. B. Boycott, E. G. Gray, and R. W. Guillery, Synaptic Structure and Its Alteration with Environmental Temperature: A Study by Light and Electron Microscopy of the Central Nervous System of Lizards, *Proceedings of the Royal Society B*, 154 (955), Plate 20, DOI: 10.1098/rspb.1961.0026 Copyright © 1961, The Royal Society.

boutons, was mysterious. Why were the fibrils more obvious in the colder environment, and what role might they possibly play in synaptic functions? We had no answers.

7.3 The structure of synapses as seen with an electron microscope

By the late 1950s, when we were finding these rings in the lizards, the electron microscopic details of synaptic structures had already been well defined (Palay and Palade 1955; De Robertis and Bennett 1955). Thanks to JZ's energy and foresight the department had a new facility for electron microscopic studies and JZ had earlier undertaken an electron microscopic study of synaptic structures on the motor neurons of the mammalian spinal ventral horn (Wykoff and Young 1956). George Gray, who had joined the department some time earlier, had

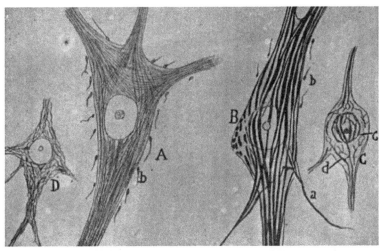

Figure 7.2 Ramón y Cajal's drawings of the fibrillary contents of cells from a lizard's brain. Those on the left show cells from a lizard that had been kept in the warm and those on the right show cells from a lizard caught in the winter in Madrid. Note cell A is also shown as Figure 5.3(b) in Chapter 5.

From Ramón y Cajal (1955).

been studying the structure of synaptic contacts in the cerebral cortex with the **electron microscope** (Gray 1959), where the reduced silver methods had shown no synaptic boutons at all. He had long urged me to move from light microscopy to electron microscopy, so we joined forces to learn more about the thermolabile fibrillar rings in the lizard brains. We found that, sure enough, these rings were in the terminal presynaptic parts of axons (Figure 7.3) and resembled the synaptic rings that had been described by light microscopists in the spinal cord and brainstem (see Chapter 5, Figure 5.3, and Figure 7.2). For me this was an introduction to viewing synaptic structures in an entirely new way: it introduced me not only to the quite new advances made by electron microscopists, but also to newly revealed functional processes involved in synaptic transmission (Fatt and Katz 1951, 1952; and see Nicholls 2007).

Whereas earlier light microscopic studies of synaptic relationships had been crucial to our views of nerve cells and their functions in terms of the reticularist and the neuronist interpretations, the electron microscope opened a quite different view, crucial for Chapter 8.

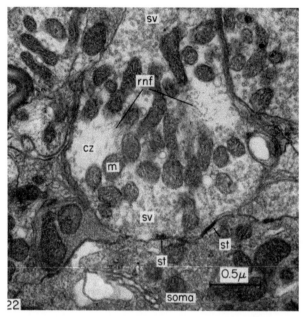

Figure 7.3 Electron microscope image of a fibrillary ring (rnf) in an axon terminal from a lizard kept in the cold. The filaments that make up the fibrillary rings are cut obliquely on the left and close to transversely on the right. They are adjacent to and lie in a zone of clear cytoplasm (cz). The terminal contains mitochondria (m) and synaptic vesicles (sv) and makes synaptic contacts (st) with small postsynaptic structures. Notice that the filaments are never close to the synaptic contacts. The gap shown by the neuronists between the fibrillar boutons and the postsynaptic cell was not the synaptic gap.

Reprinted from *International Review of Cytology*, 19, E.G. Gray and R.W. Guillery, Synaptic Morphology in the Normal and Degenerating Nervous System, p. 139, Figure 22, doi.org/10.1016/S0074-7696(08)60566-5 Copyright © 1966 Academic Press Inc. Published by Elsevier Inc. All rights reserved.

When synaptic junctions were first studied with the electron micro-scope by Palay and Palade (1955) and by De Robertis and Bennett (1955) (see Figure 7.4), it was evident that there was a clear gap between the presynaptic and the postsynaptic membranes. This was generally cel-ebrated as a significant confirmation of the neuron doctrine, although it was not widely recognized at the time that the very narrow gap (see 'sc' in Figure 7.4(d)) was only about 120 Ångstroms (= 0.012μm), far below the resolving power of light microscopes then available, which was about 0.5μm (0.5μm = 0.0005mm). While the electron microscope seemed to be supporting the neuronists by demonstrating a synaptic

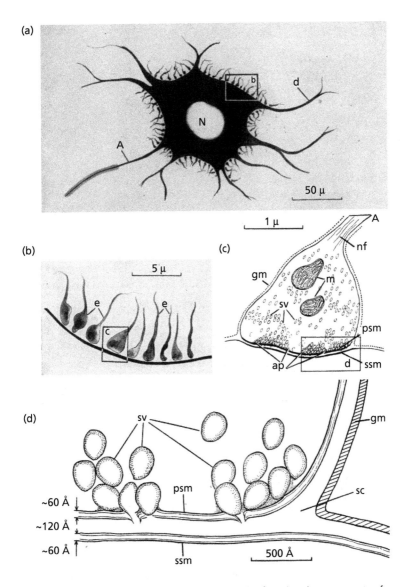

Figure 7.4 Figure to illustrate the distribution and the functional components of synapses.

(a) and (b) Representations of the light microscopic appearance.

(c) and (d) The electron microscopic appearance of the synapses. A, axon; ap, synaptic thickening; d, dendrite; e, axonal endings; gm, membrane of a supporting glial cell, which lies close to the synaptic junctions. These commonly lie immediately adjacent to a synaptic junction and play a crucial role removing some of the products of synaptic activity; m, mitochondria, these provide a source of energy for synaptic transmission; nf, neurofilaments; psm, presynaptic membrane; sc, this marks the junction of the synaptic gap with the extracellular space that surrounds the axon terminal; ssm, subsynaptic membrane; sv, synaptic vesicles.

Reprinted from *International Review of Cytology*, 8, Eduardo De Robertis, Submicroscopic Morphology of the Synapse, pp. 61–96, doi.org/10.1016/S0074-7696(08)62728-X Copyright © 1959 Academic Press Inc. Published by Elsevier Inc. All rights reserved.

gap, it also appeared to be supporting the reticularists by confirming that no real synaptic gap could possibly have been seen by the neuronists.[58]

We saw in Chapter 5 in Figure 5.5, that the reticularists had illustrated nerve cells with a far richer covering of incoming axon terminals than any shown by the neuronists (see also Held 1906). Held's and Auerbach's figures of a dense synaptic covering of nerve cells were dismissed by neuronists as artefacts produced by the method; wrongly, as was shown later by the electron microscope. Wykoff and Young in 1956, had expressed surprise at the rich covering of synaptic contacts seen with the electron microscope on cells of the spinal cord, and in view of this observation Armstrong, Richardson, and Young (1956) reproduced Auerbach's preparations by a method designed to stain the many **mitochondria** that characterize electron microscopic images of synaptic terminals (labelled m in Figure 7.4(c)[59]). Armstrong and colleagues confirmed the accuracy of Auerbach's figure, particularly his interpretation concerning the density of the synaptic covering. The history concerning synaptic numbers and synaptic gaps was confusing at the time. Once the details of synaptic structures had been understood in terms of the electron microscopic images, the earlier detailed descriptions, interpretations, and fierce arguments about the light microscopical images became comprehensible, but also, rather suddenly, became irrelevant. They now represent a piece of history that illustrates how difficult it often was to arrive at a clear view of neural functions. For us at the time, the electron microscope represented the beginning of a major change in the way neuronal interactions could be understood.[60]

---

[58] This is a point that made Ramón y Cajal's argument about the Golgi stain not crossing from one cell to another at synaptic junctions important. The subject moved ahead successfully, leaving the confusions of the past essentially unresolved and rarely discussed. That is the way science progresses; why brood about past failures and muddles?

[59] Although only two mitochondria are shown in this figure, the section is extremely thin and the much thicker sections used for light microscopy would generally show a small crowd of mitochondria.

[60] There was one other important change that occurred at about this time. Photographic images replaced drawings almost completely, no matter what type of microscope was

So far as the fibrillary rings stained by reduced silver methods were concerned, the electron microscope showed nothing that seemed to make them relevant for understanding synaptic functions. Some synapses have them, others lack them. The rings seen in the lizards resembled similar rings commonly seen in the mammalian spinal cord and brainstem. Although earlier investigators had shown that they tended to increase when an axon was degenerating (Figure 7.5) (Hoff and Hoff 1934), they rarely came close to the actual synaptic contacts (labelled st in Figure 7.3) and commonly they were absent from synapses altogether. In the cerebral cortex, the synapses had been readily demonstrated with the electron microscope by George Gray (1959) and later by many others, but usually these cortical synapses showed no evidence for presynaptic fibrillary contents and none of these synaptic junctions had earlier been visible to light microscopists.

7.4 The functional importance of the electron microscopic observations

The electron microscope changed our view of synaptic structures and their functional significance. The focus was now on the structures that clearly related to synaptic transmission (see Figure 7.4). The **synaptic vesicles** shown in the figure proved to be small membrane-bound packets of **neurotransmitters**, the molecules that are released from the axon terminals when an action potential reaches the terminal. The neurotransmitter can then act on the membrane of the postsynaptic cell to change the electric potential across the membrane, the membrane potential. The presynaptic membrane thickenings at the actual contact region are formed by proteins that play a role in fusing the vesicles to the membrane when the action potential reaches the terminal membrane. This fusion opens the vesicles that are next to the membrane and releases the **transmitter** molecules into the synaptic gap.

---

being used. Drawings of electron microscopic relationships such as that of Figure 7.4 are now relatively rare. These microscopes came with the cameras attached! The very personal characteristics of the nineteenth-century drawings were completely lost in most of the later publications.

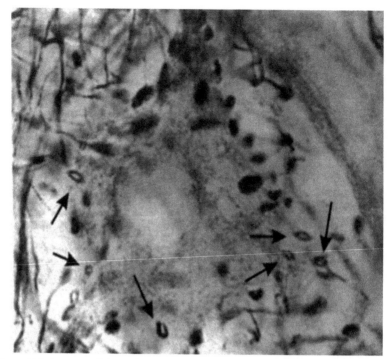

Figure 7.5 A nerve cell from the sixth thoracic segment of the spinal cord of a cat 3 days after the fifth to seventh dorsal roots had been cut. The figure shows several normal, ring-shaped axon terminals in the lower parts of the figure (at arrows) and enlarged, darkly stained degenerating terminal structures in the upper parts.

This material was originally published in E. C. Hoff and H. E. Hoff, Spinal Terminations of the Projection Fibres from the Motor Cortex of Primates, *Brain*, 57 (4), pp. 454–474, Figure 15, Plate XV, http://dx.doi.org/10.1093/brain/57.4.454 © Oxford University Press, 1934 and has been reproduced by permission of Oxford University Press.

The postsynaptic membrane thickenings represent (among others) the molecules that act as 'receptors', molecules upon which the transmitters act to produce (directly or indirectly) a change in the postsynaptic membrane potential. This change, depending on the transmitter, decreases or increases the potential across the local membrane of the postsynaptic cell, making the cell more or less likely to fire an action potential itself.

Although there are many neurotransmitters, two play key roles from the point of view of the rest of this book: glutamate depolarizes the membrane, acting to excite the postsynaptic cell, and **GABA** (gamma aminobutyric acid) in the adult brain most commonly hyperpolarizes the postsynaptic membrane and inhibits the activity of the postsynaptic cell. The mitochondria (labelled m in Figures 7.3 and 7.4(c)) provide the energy needed for synaptic transmission.

My interest in the fibrillary or filamentous[61] contents of the synaptic boutons seemed to be a red herring, irrelevant for understanding synaptic transmission and now ready for a rubbish heap of historical remnants together with the evidence about the synaptic gap and the synaptic numbers. Fortunately, I did not understand this at the time. I kept my interest in the filamentous contents of axons and their terminals and started to study the axons that carry messages from the retina to the thalamus. I chose these because an earlier study by Glees and Le Gros Clark (1941) at the University of Oxford studying the visual pathways of monkeys, had shown that the **retinogeniculate axons**, when cut, showed a remarkable increase in the fibrillary contents of their terminals (see Figure 7.6), far more dramatic than anything reported earlier (e.g. Figure 7.5). The changes described by Glees and Le Gros Clark seemed unusual and attracted my attention because this pathway was already well recognized for showing an unusual reaction when the retinogeniculate axons had been cut: the geniculate cells in receipt of the cut retinal input were known to show a remarkable 'transneuronal' (better described as a trans-synaptic) cell shrinkage. Glees and Le Gros Clark interpreted their observations as showing that each geniculate cell body was contacted by one of these giant degenerating axons (see Figure 7.6(b)) and it looked as though the structural dependence of the geniculate cell on this one enormous retinal input might well underlie the somewhat unusual shrinkage of the denervated geniculate cells.

In order to find out more about a possible link between the unusually large accumulation of neurofibrils in the terminals of these cut retinogeniculate axons and the shrinkage of the geniculate cells, Marc Colonnier, a Canadian doctoral student in the anatomy department at University College London, and I started to study these degenerative changes with the electron microscope. Marc was just finishing his thesis

---

[61] Fibrils refer to the structures seen by light microscopy and filaments refer to the much finer structures seen by electron microscopy.

(a)

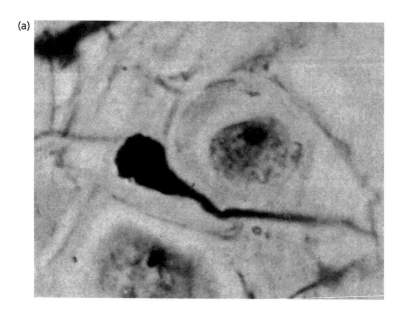

(b)

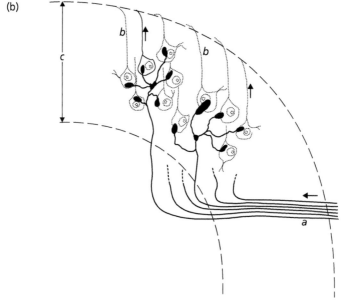

Figure 7.6 Degenerating retinogeniculate axon terminals from the lateral geniculate nucleus of a monkey.

(a) A grossly enlarged terminal of a degenerating retinogeniculate axon that had earlier been cut. Note its close relation to the cell body.

(b) The proposed one-to-one relationship between such large axon terminals and the individual geniculate cells.

Reproduced from P. Glees and W. E. le Gros Clark, The termination of optic fibres in the lateral geniculate body of the monkey, *Journal of Anatomy*, 75 (3), p. 301, Figures 5 and 6 © 1941, Anatomical Society.

research of degenerative changes in the cerebral cortex, where the synaptic terminals contained no filaments. They did not swell significantly when their axons had been cut, but rapidly became engulfed within the **cytoplasm** of supporting cells, digested together with only a tiny part of the postsynaptic cell, leaving the main part of the postsynaptic cell intact and not showing any trans-synaptic degeneration. We thought that the large geniculate terminals might carry correspondingly large parts of the postsynaptic cells to be engulfed and digested by glial cells, thus perhaps explaining the transneuronal degeneration. We were proved wrong, but it was the large degenerating boutons in the lateral geniculate nucleus that would lead me to spend most of my career studying the thalamus. This is the focus of Part III of this book.

Part III

Arriving at the thalamic gate

Chapter 8

Defining the functional components of the thalamic gate

8.1 Summary

This chapter is heavily based on Sherman and Guillery (2013; and see Sherman 2016). It starts by summarizing the electron microscopic appearance of the retinogeniculate axons and their immediate environment. These form the functional components of the visual input to the thalamic gate. The evidence is then presented for treating all major thalamic relay nuclei as having a shared structure, produced by a shared developmental and evolutionary origin. Each nucleus receives a small proportion of its synaptic inputs (<10%) for relay to the cortex; these are the *drivers*. Drivers are topographically organized with the topography representing body parts, sensory space, or parts of the brain. Some drivers come from sensory pathways or from subcortical regions of the brain, and these innervate *first-order* thalamic relays; another, major part of the thalamus receives its drivers from the cerebral cortex itself, and these form the *higher-order* relays to the cortex. These higher-order corticothalamic inputs are crucial for understanding cortical processing.

A large proportion of synaptic inputs (>90%) are not relayed to the cortex and are classifiable as *modulators*. They contribute to controlling the gate. Some modulators match the topography of the drivers, thus relating to the parts of the body and the world; others do not show this specificity and generally have more global actions.

8.2 Synaptic terminals in the lateral geniculate nucleus

The lateral geniculate nucleus commonly has six layers in macaque monkeys and humans: three layers receive their inputs from the left eye and three from the right eye (Figure 8.1). In our experiments, Marc Colonnier

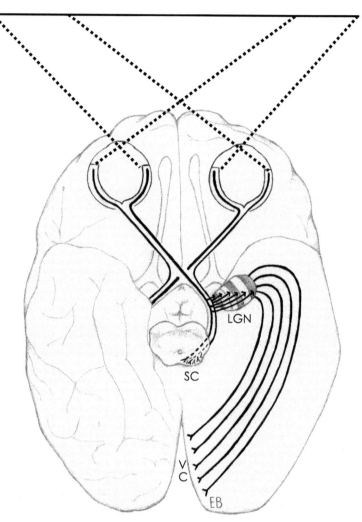

Figure 8.1 A schematic view of the major visual pathways as seen in relation to a ventral view of the cerebral hemispheres (from Figure 1.1(c)). The retina is divided into two unequal parts, the larger lying nearer the nose and sending more axons to the brain (thicker line). Some of the axons from each eye cross at the midline so that the left hemisphere receives from the right part of the visual world and the right hemisphere receives from the left. Distinct functional components of the retinofugal pathways go to distinct layers of the lateral geniculate nucleus (LGN) before these separate messages are all relayed to the visual cortex (SC). In addition to this, many, possibly all, of the axons that go to the lateral geniculate nucleus send branches to the superior colliculus or to the pretectal nucleus. Only the former are shown in the figure.

Illustration by Dr Lizzie Burns.

and I cut one optic nerve to reproduce the degeneration of the retino-geniculate axons demonstrated by Glees and Le Gros Clark; we used the normally innervated layers as a control for the denervated layers.

Our results confirmed one aspect of the Glees and Le Gros Clark study: very large processes densely filled with filaments were present in the denervated layers (Figure 8.2(a, b)) but not in the normal layers (Figure 8.2(a)). Whereas at early stages (5 days' survival, Figure 8.2) some were barely distinguishable from filamentous rings in the normal layers, others were densely filled with filaments (Figure 8.2(b)). At later stages none looked normal. By 56 days no recognizable retinogeniculate terminals remained: they had all been eaten up by local or invading scavenger or glial cells. Figure 8.2 shows that the large filamentous

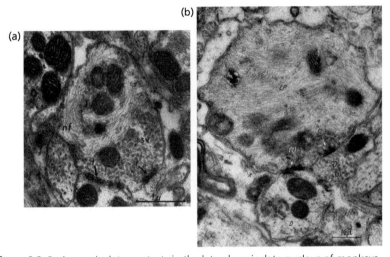

Figure 8.2 Retinogeniculate contacts in the lateral geniculate nucleus of monkeys.

Parts (a) and (b) show large degenerating retinogeniculate axon terminals 5 days after section of an optic nerve in a monkey. Both are filled with a dense population of neurofilaments (nf in (a)); these correspond to the neurofibrils identifiable by light microscopically (see Chapter 7). This increase in filaments is far more advanced in (a) than in (b). In the lower part of each figure two synaptic junctions are shown. In (b), the synaptic contacts establish a 'triad' at arrows, 1, 2, and 3. The vesicle-filled (SD) middle profile represents a presynaptic dendrite. Better examples of triads are present in Figure 8.3. Further details in the text.

Reproduced from *Zeitschrift für Zellforschung und Mikroskopische Anatomie*, Synaptic organization in the lateral geniculate nucleus of the monkey, 62 (3), pp 333–355, Figure 13, DOI: 10.1007/BF00339284, M. Colonnier and R. W. Guillery © Springer-Verlag 1964. With permission of Springer.

structures were, indeed, lying within characteristic synaptic terminals, with mitochondria and synaptic vesicles close to a patch of thickened synaptic membrane. The retinal inputs could be characterized not only by the fibrillar changes, but also by their structure and synaptic relationships, so that eventually even in the normal layers retinal terminals could be identified (Figure 8.3).

Two features provided intriguing puzzles for further studies: one was that the postsynaptic structures were not the nerve cell bodies required for the simple one-to-one relationship illustrated by Glees and Le Gros Clark (Figure 7.6(b)). Instead, the postsynaptic structures were all smaller profiles, sometimes recognizable as dendrites. In addition, a great many synapses were established by presynaptic terminals that never showed any degenerative changes even 56 days after the cut. That meant they were not coming from the retina and that we needed to define their origins and functions.

We knew that thalamic nuclei contained two distinct cell types: relay cells and local '**interneurons**' (see Chapter 5, Figure 5.2(a) and (b), respectively). Relay cells send their axons to the cortex whereas interneurons that have axons[62] all terminate close to the cell body. These local axons would be expected to survive our experimental cuts. There was also evidence that the cortex sends axons back to the thalamus and these, too, should survive in our experiments.

Some of the surviving postsynaptic profiles were strange because they were postsynaptic to a large retinal terminal which contacted the dendrite of a geniculate cell and were themselves *also presynaptic* to this same dendrite (see Figure 8.2(a) and Figure 8.3, at arrows). That is, they were the middle profiles of 'serial synapses', later called **triads**.[63]

---

[62] Many interneurons appear to lack axons. They are discussed later in this chapter.

[63] At the time, the triads presented a strange new problem because the middle process was both presynaptic, like an axon according to the law of dynamic polarization, and also postsynaptic, like a dendrite. It had been shown a little earlier, that in the retina, the **olfactory bulb**, and in other parts of the brain of several species, dendrites could form such serial synapses (Kidd 1962; Rall et al. 1966). Often the relevant cell lacked an axon. The 'law' of dynamic polarization had been accepted by many for decades and had proved very useful. It had survived as a law because it was useful as far as it went for the majority of known nerve cells that have axons. In the absence of a better law, the only replacement

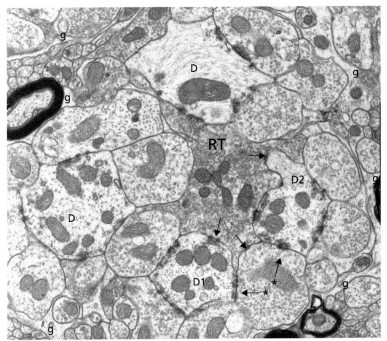

Figure 8.3 A section through a large glomerulus from the lateral geniculate nucleus of a normal cat. A part of a retinal terminal (RT) is shown centrally forming synaptic contacts with several dendrites (D, D1, D2) that contain neurofilaments, a common characteristic of dendrites close to their primary branch points near the nerve cell body. Two arrows within the retinal terminal show synaptic contacts made onto dendrites D1 and D2; another shows a contact onto a vesicle-filled local presynaptic dendritic terminal. Two arrows marked by asterisks each show one of the many vesicle-filled presynaptic dendritic terminals completing two triadic junctions. Glial processes (g) are characteristically almost completely absent from the glomerulus but lie adjacent to it.

Reproduced from S. Murray Sherman, and R. W. Guillery, *Exploring the Thalamus and Its Role in Cortical Function*, Second Edition, Figure 3.6, © 2006 Massachusetts Institute of Technology, by permission of The MIT Press.

for the law, during a long period of more than 50 years, was ignorance about the few cells that had no axon and thus seemed to have no outputs; neuroscientists kept the law and these strange cells with no axon remained in a functional limbo for decades: no one thought that the law of dynamic polarization had been 'falsified'. It was treated as a generalization with a few strange exceptions. These discrepancies between accepted theories, laws, or hypotheses, like the confusions about the synaptic gap and synaptic numbers mentioned in Chapter 7, were never resolved, they simply ceased to be of interest as the functional organization of synapses started to be understood in terms of the structures revealed by the electron microscope.

For us, these triads served as one landmark for identifying the incoming retinal axons in a normal brain, as also did a complicated arrangement of synaptic profiles that had previously been called a **glomerulus** (Szentágothai 1963), which were a clearly identifiable region (Figure 8.3), containing retinal terminals and triads. The glomeruli were distinct from neighbouring regions because they lacked the processes of supporting glial cells that in most parts of the brain lie closely adjacent to each synapse.[64]

We concluded that (1) the simple one-to-one relationship between retinal inputs and geniculate cells, proposed by Glees and Le Gros Clark does not exist;[65] (2) the retinal inputs represent but a small minority of the synapses in the lateral geniculate nucleus;[66] and (3) the synapses formed by the retinal inputs are readily recognizable by their appearance, their size, their multiple synaptic contacts, and their relationship to triads and glomeruli. We needed to learn what the other inputs in the nucleus contribute to the relay. Above all, we needed to learn whether the visual relay we had been studying was representative of thalamic relays in general.

The intermediate components of the triads in the thalamus were later shown to be dendrites of the local inhibitory interneurons (see Chapter 5, Figure 5.2(b)) that do not send axons to the cortex (Ralston 1971; Pasik et al. 1973; Hamori et al. 1974). The triads contribute to the complexity of the thalamic relay and provide a significant feature for identifying axons that carry messages for relay to the cortex. They were shown to be inhibitory, and may serve to keep the firing of the thalamic cells within their middle range so that increases and decreases in firing can always be relayed, rather than the actual rate of the input.

The interneurons play no further role in this book except for the curious fact that they are absent in almost all of the thalamic relay nuclei of mice and rats other than the lateral geniculate nucleus, which has them in all species studied. So far this makes no sense at all, but (understandably) has not stopped people from studying the thalamus of mice and rats, since these animals are cheaper and breed more readily than cats and monkeys.

[64] The supporting glial cells (labelled gm in Figure 7.4 in Chapter 7) are important for many synapses, removing potassium ions and surplus transmitters produced by synaptic activity from the extracellular spaces. The functional significance of this unusual glial relationship for some thalamic synapses has not been clearly defined, although it could also contribute to preventing high firing rates, keeping firing rates at intermediate levels relaying changes in signal strength rather relaying actual stimulus strength.

[65] This had earlier been demonstrated by Szentágothai (1963).

[66] Quantified later by Van Horn et al. (2000) and van Horn and Sherman (2007).

The thalamic relay was clearly far more complicated than we, or anyone else had anticipated. In Chapter 1, I compared our view of the thalamic relay with the contents of a Victorian novel; now we seemed to be at the title page of a rather thick novel, much of which still remains to be read (and understood) even today.

8.3 Generalizing from one thalamic nucleus to others on the basis of their shared developmental history

It seemed reasonable to expect similar synaptic structures in other thalamic nuclei. Earlier embryological and evolutionary studies had shown that each major brain part has a distinctive developmental history and, consequently, each has a characteristic, basic organizational plan, each plan differing from those of other major brain parts. The spinal cord and brainstem share a common structure even where different parts have distinct functions and the same is true of the six-layered cortex and also of the different parts of the cerebellum.

In a comparable way, the thalamus has a number of separate nuclei (see Chapter 1, Figure 1.1(d), and Figure 8.4), each receiving its major inputs from a different pathway, each sending outputs to a distinctive cortical area or group of areas, and each serving different functions.[67] In terms of what is known today about the structural organization of the thalamus, there are many shared features as well as some features that distinguish one nucleus from another.[68]

Observations of the lateral geniculate nucleus and of other first-order thalamic nuclei in the 1960s and 1970s (see Jones and Powell 1969a,

---

[67] Many thalamic nuclei also connect to the striatum (see Chapter 4, Figure 4.1), but since this book is primarily concerned with the thalamus and cortex I include little about these connections.

[68] There is something very useful about this approach to understanding the organization of the different brain parts, but there is also, as will be obvious, something slightly dubious about it, because almost any general statement about the structure of one of the major brain parts comes with exceptions. This is something that neuroscientists have learnt to live with. Perhaps one day we will have a more powerful generalization but at present we have to define the general pattern as best we can and look for specific details if we want to understand particular nuclei or areas in terms of their functions and connections.

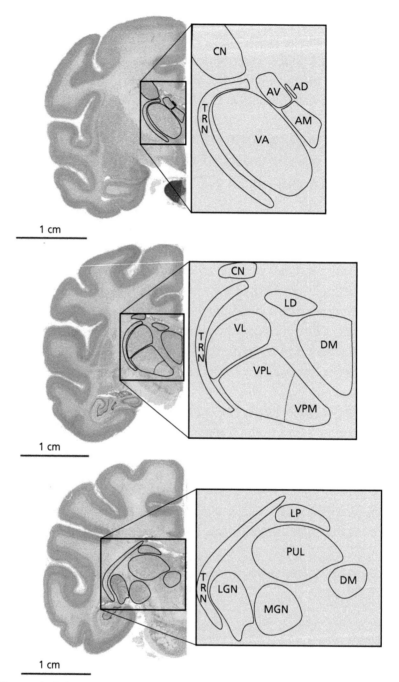

Figure 8.4 Thalamic nuclei shown in relation to coronal (a vertical cut perpendicular to the midline plane) sections through the whole brain of a rhesus monkey.

1969b; Ralston and Herman 1969; Jones and Rockel 1971; Harding 1973; Majorossy and Kiss 1976; Robson and Hall 1977a, 1977b; Mason et al. 1996; Jones 2007; and see Sherman and Guillery 2006, 2013), gradually led to a view of thalamocortical connections upon which the rest of this book is based, that the major thalamic nuclei have a shared organizational plan and receive a relatively small proportion of driver afferents for relay to the cerebral cortex.

8.4 **Identifying the drivers in the thalamic nuclei**

As different thalamic nuclei on the sensory pathways to the cortex (visual, auditory, somatosensory, etc.) were studied with the electron microscope during the 1960s and 1970s in several different species, primarily cats and monkeys, each nucleus showed the basic structural organization seen in the lateral geniculate nucleus. The inputs that had been physiologically identified as the source of the messages for relay to the cortex, the drivers, shared many of the structural features of the retinogeniculate terminals. They were large, made many synaptic contacts, including triads, were significantly outnumbered by other presynaptic terminals (Van Horn et al. 2000; Van Horn and Sherman 2007), and were commonly found in glomeruli (Figure 8.3). The same was found for the **ventral lateral nucleus** and the anterior thalamic

---

Figure 8.4 Continued

The major first-order nuclei discussed in the text are AD, anterodorsal nucleus; AM, anteromedial nucleus; AV, anteroventral nucleus (three recipients of mamillothalamic driver afferents, see Chapter 6); LGN, lateral geniculate nucleus (visual driver afferents); MGN, medial geniculate nucleus (auditory driver afferents); VL, ventral lateral nucleus (driver afferents from the cerebellum); VPL, ventral posterior lateral nucleus (somatosensory relay for body and arms); and VPM, ventral posterior medial nucleus (somatosensory relay for head). The higher-order nuclei shown are LD, laterodorsal nucleus (driver afferents from cortex receiving from AD, AM, AV); LP and PUL, lateral posterior and pulvinar nuclei (recipients of cortical driver afferents related to visual and somatosensory relays); DM, dorsal medial nucleus (driver afferents from frontal cortex); and VA, ventral anterior nucleus (driver afferents from areas of motor cortex). TRN, the thalamic reticular nucleus (see text) and CN, the caudate nucleus, are not relays to cortex.

nuclei, which relay cerebellar and mamillary messages to the cortex, respectively. These readily identifiable inputs for relay to the cortex, the drivers, needed to be distinguished from the majority of other synaptic terminals, which appeared not to carry messages for relay to the cortex, the modulators (Sherman and Guillery 1998).

Two further issues then arose: one was to identify the drivers for the many other thalamic nuclei, for example, the dorsal medial nucleus, or the pulvinar (Figure 8.4), for which we had no clear evidence concerning an origin of any drivers. The other was to learn more about the modulators

8.5 **First-order and higher-order thalamic relays**

There were many thalamic nuclei for which no driver inputs had been defined, although it was known that these nuclei all send axons to the cortex (e.g. Walker 1938). They had often been considered to be 'association nuclei', linking different sensory modalities to each other, but no specific connections had been defined for such a function.

Once the characteristic thalamic structures of drivers were identifiable, it became evident that driver inputs to these association nuclei were coming from the cerebral cortex itself (Mathers 1972; Robson and Hall 1977a, 1977b; Ogren and Hendrickson 1979; Wilson and Hendricksen 1981; Schwartz et al. 1991; Rockland 1996; Guillery 1995; Guillery et al. 2001). These driver inputs were shown to come from relatively large cells in cortical layer 5, which is the cortical layer that also represents the major output from the cortex to the several different subcortical regions illustrated in Chapter 4 in Figure 4.1, including the superior colliculus, basal ganglia, and pons, all parts of the motor apparatus of the phylogenetically ancient vertebrate brain (see Robertson et al. 2014; Grillner and Robertson 2016).

In contrast to the sensory, the mamillary, and the cerebellar transthalamic pathways, which transmit information to the cortex from non-cortical sources, these corticothalamic drivers relay messages from one cortical area to another, higher, cortical area.[69] We called the former

---

[69] So far all of the available evidence shows the connections as feedforward going from lower levels to higher levels.

first-order relays and the latter higher-order relays (Guillery 1995; Sherman and Guillery 2006). Higher-order relays have been demonstrated for former 'association nuclei' including the dorsal medial nucleus and the pulvinar. At the time, this feedforward transthalamic corticocortical pathway formed a novel set of corticothalamocortical connections, functionally clearly distinct from the direct corticocortical connections that have been far better defined in terms of details (Van Essen and Andersen 1990; Van Essen et al. 1992).

Whereas the first-order nuclei receive no cortical driver inputs for relay to the cortex, the higher-order nuclei can also receive some non-cortical driver inputs. For this reason, it is necessary to distinguish first- and higher-order nuclei from first- and higher-order relays.

For a few higher-order nuclei, specifically for the visual pathways of cat and monkey and the somatosensory and auditory pathways of the mouse, physiological and morphological evidence has shown that, indeed, thalamocortical links are demonstrable for the proposed cortical drivers to the higher-order nuclei (Bender 1983; Chalupa and Abramson 1988; Chalupa 1991; Diamond et al. 1992; Petrof and Sherman 2013).

The transthalamic pathways provide an alternative link between cortical areas. The extent to which these two sets of connections, direct and transthalamic, are parallel to each other, providing a comparable series of six or more hierarchical levels in monkeys and almost certainly more levels in the human brain, remains to be determined, as do the different functional roles of the two systems. So far, we only have evidence about the lowest levels of the transthalamic connections, and these look as though they are running parallel to the direct pathways. This currently represents an important area of ignorance for anyone hoping to understand thalamocortical relationships.

8.6 **The modulators**

It was clear that many of the inputs to the thalamus were not carrying messages for relay to the cortex.[70] By the 1990s, Murray Sherman and

---

[70] Most strikingly, the cells in layer 6 of the visual cortex respond to visual stimuli in a pattern quite distinct from the pattern seen in the retinogeniculate axons or in the cells of the

I had been interacting in our studies of the visual pathways for some years and during the 1990s we started to focus our attention particularly on the lateral geniculate nucleus as an example of thalamic organization in general.[71] Murray's background had included a spell in PO Bishop's group, one of the most rigorous groups of visual neurophysiologists at the time, and his studies had included a far more functional approach to thalamic organization than mine. He had also produced detailed quantitative electron microscopical studies of the lateral geniculate nucleus.

It was one thing to distinguish drivers from modulators on the basis of whether or not they relayed messages to the cortex, but Murray also focused on defining the functional features that distinguished these two inputs.

The drivers in the thalamus all use glutamate as a transmitter. Some modulators also use glutamate but a number of other transmitters also serve as modulators. The drivers all have a rapid action, producing postsynaptic membrane changes that only last a few milliseconds, whereas the modulators produce much longer postsynaptic changes lasting up to a second or longer (see Sherman 2014). This difference depends partly on the receptor molecules that lie in the postsynaptic membrane. The transmitter can either act directly, and very rapidly, on the ion channels (**ionotropic receptors**) in the postsynaptic membrane or act more slowly through a chain of reactions that eventually opens or closes the ion channels (metabotropic receptors). Some **glutamatergic** synapses, the drivers, lack metabotropic receptors and can transmit messages rapidly to communicate rapid actions happening in the world or the body. Others have metabotropic receptors and must thus act more slowly.

The drivers generally contact the postsynaptic cell close to the **initial segment** of the axon, where the action potential can be generated

---

lateral geniculate nucleus itself. This characteristic response pattern of the 'complex' layer 6 cells described by Hubel and Wiesel (1961, 1962) is never seen in the lateral geniculate cells. That is, the great many relatively weak layer 6 inputs to the thalamus lack the properties of drivers.

[71] Leading to the publication of Sherman and Guillery (2001) on the thalamus.

rapidly and powerfully, whereas the modulators are commonly on the peripheral dendrites far from the spike generation. One other feature concerns the pattern of repeated action potentials: drivers send a strong first action potential followed by a sequence of gradually weaker signals, whereas modulators start weak and gradually strengthen (for more details see Sherman and Guillery (2013)). These features all act to provide the drivers with a strong and rapid influence on the postsynaptic cell, features that are lacking in the modulators.

The speed of the reaction is crucial for following rapid actions in the body or the world, such as returning a service in tennis or quickly reading a line of print. Especially for the sensory pathways these are crucial features for communicating finer temporal details.

GABAergic inhibitory inputs can be regarded as modulatory components. They act to decrease the probability that postsynaptic action potentials will be generated, so they can only generate a new message by eliminating action potentials. Smith and Sherman (2002) have argued that while in theory it would be possible to generate a message if the recipient cells were firing at very high frequencies,[72] the firing rates of thalamic cells are not high enough for this. They concluded that the GABAergic inputs to the thalamus cannot generate a new message in a thalamic relay cell. The GABAergic inputs will here be treated as modulators of thalamic activity that decrease the chances of a message being relayed.

8.6.1 The reciprocal corticothalamic pathway from cortical layer 6

In addition to the layer 5 neurons that provide drivers for higher-order relays, each thalamic nucleus also receives a great many excitatory, glutamatergic inputs from cells in cortical layer 6. These thalamic inputs act as modulators and are not further relayed to the cortex. They, like the tiny inhibitory interneurons, have a topographically limited action because they provide a reciprocal corticothalamic connection back to the zone of the thalamus that innervates the cortical area in which they lie (see Figure 8.5(a)). That is, the cortical area damaged to produce the

---

[72] Rather like sculpting a figure from a block of stone; removing action potentials from a block of very closely placed action potentials.

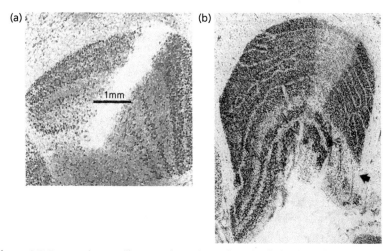

Figure 8.5 Two sections to illustrate the reciprocal corticothalamic and thalamocortical connections.

(a) Section through the lateral geniculate nucleus of an owl monkey showing a pencil of radioactive label passing through all of the layers of the nucleus. A small area of the visual cortex had been labelled with tritiated proline, and this had been transported through the axons of the layer 6 corticogeniculate axons to the lateral geniculate nucleus. A photographic emulsion over the surface of the geniculate sections shows the localized distribution of the radioactive terminals in the section of the nucleus.

Reproduced from S. Murray Sherman, and R.W. Guillery, *Exploring the Thalamus and Its Role in Cortical Function*, Second Edition, Figure 3.8, © 2006 Massachusetts Institute of Technology, by permission of The MIT Press.

(b) Photograph of a pale sector of retrograde degeneration in the lateral geniculate nucleus of a mandrill (*Mandrillus sphinx*). This zone of cell loss was produced by a limited zone of damage to the visual cortex. The geniculate cells died because their axons had been damaged at their cortical terminals. The zone of cortical damage and the zone of geniculate cell loss both correspond to a small region of the retina from which they received their visual inputs. Note the accurate border of the zone of cell loss. The lines defining this border in the figure correspond to points in the retina.

Reproduced from J.H. Kaas, R.W. Guillery, and J.M. Allman, Some Principles of Organization in the Dorsal Lateral Geniculate Nucleus, *Brain, Behavior, and Evolution*, 6 (1), p. 358, Figure 3, DOI:10.1159/000123713 © S. Karger AG, 1972.

cellular degeneration shown in Figure 8.5(b) would in a normal animal include layer 6 cells that could produce labelling like that in Figure 8.5(a) in the same region that shows cellular losses in Figure 8.5(b).

In the cortex, the layer 6 cells significantly outnumber the layer 5 cells and in the thalamus, the layer 6 cell synaptic junctions greatly outnumber the synaptic junctions made by drivers. In major first-order thalamic nuclei for which numbers are available, the layer 6 inputs represent about 30% of the synaptic junctions[73] compared to less than 10% for the drivers. We need to think about functions for this large population of topographically focused excitatory inputs acting against a correspondingly local group of inhibitory interneurons.

8.6.2 The thalamic reticular nucleus

A thin shell of cells, lateral and dorsal to the main part of the thalamus, called the thalamic reticular nucleus (labelled TRN in Figure 8.4) provides an additional, localized inhibitory input to thalamic relay cells. The reticular cells are developmentally distinct from the rest of the thalamus and have a completely different organizational plan. The reticular cells lie on the pathways of the topographically related thalamocortical and layer 6 corticothalamic axons (see Figure 8.6) receiving glutamatergic excitatory inputs from both. The reticular cells, in turn, send inhibitory, GABAergic inputs to the same segment of the thalamic nucleus from which they receive excitatory inputs, allowing them to influence the balance of excitation and inhibition acting on the topographically related relay cells (Montero et al. 1977; Crabtree 1992, 1996, 1998). These tight topographical links led Crick (1984) to propose that the corticoreticular connections could serve like a searchlight focusing on particular segments of the thalamocortical pathways.

8.6.3 Other modulators

Several modulators in the thalamus represent systems that are not specific to the thalamus and show a limited topography in relation to the parts of a thalamic nucleus. They innervate many parts of the brain and their activity relates to the general state of the organism, such as a feeling of well-being, chronic pain, hunger or thirst, or a general state of alertness, which generally do not require rapid and brief signals.[74]

---

[73] Observations on cat lateral geniculate nucleus (Erişir et al. 1997; Van Horn et al. 2000).

[74] These include cholinergic, adrenergic, serotonergic, and histaminergic components that play no further role in this book.

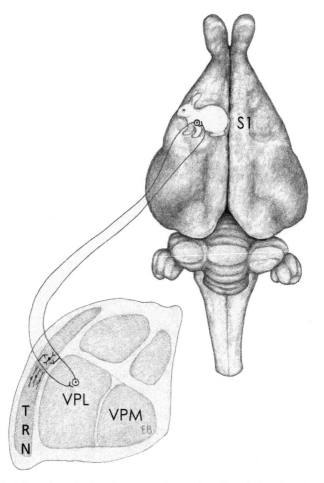

Figure 8.6 A figure based primarily on experimental studies of the rabbit (Crabtree 1992) showing the pattern of the connections formed between the somatosensory cortex (S1), the posterolateral thalamic nucleus (VPL and VPM), and the thalamic reticular nucleus (TRN). Further details in the text.

Illustration by Dr Lizzie Burns.

Additional GABAergic inhibitory inputs come from regions of the brain that otherwise play a minor or no role in this book. They are best considered as linked to motor control systems that receive significant inputs from the cortex.[75] We do not understand the details of their

---

[75] These include the **globus pallidus**, the substantia nigra, the zona incerta, and the **pretectal nucleus**; see Lavallée et al. (2005) and Chapter 3 of Sherman and Guillery (2013).

specific role in the control of the thalamic gate. There is no clear evidence about their topography and no good reason to think of them as topographically organized to match the connections of the cortical and reticular cells.

However, there is one particularly interesting feature about these inputs: they go predominantly to the higher-order nuclei, indicating that they are probably able to shut down higher-order relays while first-order relays continue to function, and may well be doing this particularly during motor actions. The possibility that first-order relays are left open while higher-order relays are inhibited can be seen as a general mechanism for allowing the cortex to have continuing access to subcortical mechanisms even when higher cortical areas are inactive (as during low levels of attention). Alternatively, or additionally, this can provide a mechanism in an awake animal for reducing conscious levels of control, leaving some (well-learnt?) motor actions to lower levels.[76]

8.7 The actions of the thalamic gate

The major function of the thalamus is to transfer the messages that it receives from the world, the body, and from other brain centres to the cortex. Elliott-Smith had referred to it as a gate, and this idea was drummed into us as students. However, it is a rather unusual gate: it is a one-way gate—it can pass on to the cortex all the information that it receives about the world, the body, and the brain and this, as indicated earlier, is essentially all of the information that the cortex receives about events in the world, the body, and the brain. The many messages that layer 5 cortical cells send back to the lower brain centres, pass by a more lateral route. They do not go through the gate; only the thalamic branches for relay to the cortex go through the gate. The thalamic gate cannot act on these more lateral motor outputs of the cortex. This is

---

[76] This arrangement should be considered in relation to the observations cited by Grillner (2015) concerning the importance of motor cortex for learning a motor skill but not for executing it. Once learnt, the lowest sensory thalamocortical levels could be sufficient for the executive functions and would also provide the actions of the thalamic gate for catching errors or irregularities.

crucial for understanding the functional organization of the transthalamic part of the cortical hierarchy in the following chapters: whereas the thalamic branches of layer 5 cells are subject to thalamic controls, their motor branches are not.

The thalamic gate for any one relay cell is not just either open or shut but has three distinct functional states which relate to the membrane potential of that relay cell and its immediate past history. These relationships are described fully in Sherman and Guillery (2013), and will only be briefly summarized here. The gate can be open, in 'tonic mode' when the cell is relatively depolarized not far from its threshold, and simply passes information on to the cortex very much in the form in which the information reaches the thalamus.[77] This response is shown in Figure 8.7(a), where the luminance of the stimulus varied sinusoidally over time, as shown at the top of the figure; or the gate can be closed, when the cell is relatively hyperpolarized, so that the relay will not pass on any of the information that it receives. The third condition, the 'burst mode' (Figure 8.7(b)), depends on the relay cell having been silent and relatively hyperpolarized for at least 100 msec and then a sufficiently strong input (that is, a driver) will produce a burst of action potentials shown in Figure 8.7(b). Such a rapid burst of action potentials does not correspond to the input in terms of its temporal pattern. Here it is important to repeat that in a sensory pathway the temporal pattern of the action potentials represents the 'meaning' of the message: it relates to the dynamics of the sensory event.

The rapid burst of action potentials that arrives at the cortex in the burst mode is a markedly different and more powerful signal than the action potentials in the tonic mode, which at other times are sending information about the body and the world along that same 'labelled line'. In the burst mode, the signal-to-noise ratio (compare Figure 8.7(b and d)) is much higher than it is in the tonic mode (compare Figure 8.7(a and c)). One can think of a burst as ideally suited to wake the cortex up to an unexpected event that needs attention. Notice

---

[77] This is the way in which the thalamic gate is most generally conceived.

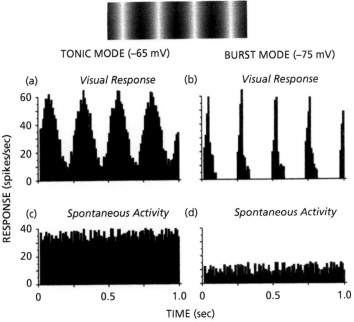

Figure 8.7 Four recordings obtained from a cell in the lateral geniculate nucleus of a cat while it was viewing a moving image of sinusoidally changing light and dark stripes as shown at the top of the figure. Part (a) shows the responses in the tonic mode following the sinusoidal changes of the visual stimulus. Part (b) shows the responses in the burst mode. Part (c) shows that the spontaneous (background, no stimulus) activity in tonic mode is fairly high allowing the response to represent the low levels of illumination as well as the high levels. Part (d) shows the spontaneous activity in burst mode. Note the higher signal-to-noise ratio ((b)/(d)) in the burst mode than in the tonic mode ((a)/(c)).

that a labelled line when acting in burst mode still carries important information about the presence of a specific event, but provides no detail about the actual nature of the events represented by the burst. That is, the message may be quite specific about the source of the event, perhaps a thermoreceptor on the middle finger of the right hand, without specifying anything specific about the strength or the time course of the stimulus. It is simply a warning signal: pay attention to the middle finger of the right hand. That is, in terms of the temporal changes of the particular peripheral events, the meaning, represented by the dynamics of the action potentials, is lost, although the cortex still has information about the origin of the message.

Since the cortex has a powerful topographically reciprocal excitatory feedback from layer 6 to the thalamic relay cells, this corticothalamic link can activate the relevant thalamic relay cells in response to an incoming burst, depolarizing the membrane of those thalamic relay cells and thus bringing them back to the tonic mode so that the thalamic cells can once more send an accurate copy of the incoming message to the cortex.

Imagine that you have been quietly reading an interesting book quite oblivious of your surroundings, the auditory pathways to your cortex are receiving minimal signals, and there is a low level of spontaneous firing from the auditory relay cells in the thalamus. Then there is the cry of a baby (your baby). The auditory cortex, which has been quiet while you were reading, receives a burst of activity. This burst is not useful information about what is happening in your auditory surroundings, but instead it stimulates corticothalamic cells that feed back to the relevant thalamic relay cells, changing the membrane potential of these thalamic cells so that they will switch to the tonic mode. Now you can quickly identify the precise nature of the noise.

So far as our present view of the thalamic relay is concerned, the many synapses that were described earlier in this chapter as modulators coming from various sources, will all play an important role in controlling the membrane potential of the relay cell, determining whether messages are passed to the cortex and whether they are passed in the tonic mode as a detailed message about ongoing events at the origin of a particular neural input or as a warning about the occurrence of unexpected events in the burst mode. Similar connections are present in all thalamic relay nuclei, although the details of the modulators vary somewhat from one nucleus to another and, in particular, higher-order nuclei receive inhibitory inputs that do not go to first-order nuclei.

If the thalamus were, indeed, just a simple relay, transmitting in the tonic state all of the time, then all of the cortex would be alert and continuously receiving a flood of messages from the body, the brain, and the world. If you were reading and the baby were not crying, you might not be able to focus on your book and ignore all other inputs. We need to think of the preponderance of modulatory synaptic inputs that reaches the thalamus as serving to control the thalamic gate, adjusting the relays to the current attentional demands.

Although the present account assumes that the only relevant variable for the action of the many afferents other than the specific driver inputs is the control of the relay cell's function as a gate on the way to the cortex, this assumption may well be wrong. Although the control of the thalamic gate is undoubtedly an important function relevant for the chapters that follow, it would be unwise to assume that no other functions, as yet undefined, play a role in the thalamic relay.

Chapter 9

Thalamic higher-order driver inputs as sensorimotor links

9.1 Summary

This chapter provides a closer look at the thalamic and motor branches of the driver inputs to higher-order thalamic nuclei, and introduces their functional significance for discussion in later chapters. The thalamic branches bring information for relay to higher cortical levels, including a copy of the information carried in the motor branches about anticipated cortical contributions to the control of actions at lower levels and consequent changes in perceptions. In this way, the higher cortical level can add to the control of an action when there is a mismatch between action and perception. Most of these branched axons that have so far been described come from early sensory areas and only a few from higher areas have been studied. We can see the hierarchy of cortical areas as providing an opportunity for higher areas to monitor lower areas and, when needed, to contribute needed control of the phylogenetically older brainstem and spinal centres. A far more extensive review of the branched thalamic driver inputs and their contributions to the control of actions than we have at present will be crucial for understanding the full complexity of this transthalamic corticocortical hierarchy.

9.2 Motor branches of higher-order thalamic driver inputs

The branching patterns of the driver inputs to first-order thalamic nuclei were considered in Chapter 2 (Figure 2.6); in this chapter and in Chapter 10, I consider evidence about similar branching patterns of inputs to higher-order nuclei, and look at their functional importance.

A great many axons from layer 5 of the cortex go to phylogenetically ancient subcortical motor centres, as shown in Figure 4.1. These centres, as indicated in Chapter 4, include not only the striatum and the pons, providing inputs to the basal ganglia and the cerebellum, respectively, but also the superior colliculus. We saw that each of these three phylogenetically ancient centres receives input from a great many different cortical areas, and that there is probably no cortical area lacking such motor outputs. Even the primary sensory areas have such outputs, which is puzzling in terms of the standard sensory-to-motor view of cortical functions.

In addition, there are several brainstem centres, some not individually shown in Figure 4.1 in Chapter 4,[78] all concerned with motor control, which also receive inputs from cortical layer 5. Many of these layer 5 corticofugal axons have thalamic branches that go to higher-order thalamic nuclei (see Deschênes et al. 1994; Bourassa et al. 1994, 1995; Rockland 1996, 1998, 2013; Veinante et al. 2000; Guillery et al. 2001; Bajo et al. 2007; Deschênes et al. 2011). It has to be stressed that the branching patterns of corticothalamic axons coming from cortical layer 5 have so far been explored to only a limited extent, hardly at all for higher levels of the cortical hierarchy of the primate brain. Once the probable role of such branches in the cortical hierarchy that monitors our ongoing conscious lives (explored in more detail in Chapter 13) is recognized, the importance of defining this system of branching axons should prove a wonderful subject for novel views of thalamic and cortical functions.

Driver afferents from cortical layer 5 cells to the thalamus for relay to the cortex have been described for visual, somatosensory, auditory, and motor cortex, in rats, cats, and monkeys. They have been shown to be branches of corticofugal axons going to the pretectum, superior colliculus, or the pons. It is reasonable to anticipate that there are many others, coming from other cortical areas, going to different thalamic nuclei, and innervating other motor centres, most probably including the striatum.

---

[78] For example, centres such as the locomotor centre (Chapter 1), the pretectum contributing to lens and pupillary control, the red nucleus, the medial pontine reticular nucleus (Chapter 6), and a respiratory centre.

Figure 9.1 provides an important and carefully recorded view of the potential power of these branched corticofugal axons. It shows eight axons coming from the primary visual cortical area, area 17, of a rat. Each has been traced to one or more nuclei of the brainstem concerned with motor actions: two pretectal areas concerned with pupillary and lens control, the superior colliculus, whose motor actions are discussed in Chapter 4, and the pons, which represents a key relay to the cerebellum and contains other motor centres. It should be noted that the standard view of sensory-to-motor relationships has no place for motor outputs from primary sensory areas.

Six of these eight axons provide a rich innervation of two higher-order thalamic nuclei: the lateral posterior nucleus and the lateral dorsal nucleus. The higher cortical areas receiving inputs from these thalamic nuclei can be expected to be receiving information concerning the activity in area 17 itself as well as information about the upcoming motor adjustments and their perceptual consequences. That is, these recipient cortical areas can monitor the expected smooth flow of sensorimotor interactions that are needed to keep the perceptions matching the anticipated actions,[79] and these cortical areas, in turn, also have the motor outputs for making needed adjustments.

These layer 5 outputs to subcortical motor centres reflect the power of all cortical areas to contribute to the ongoing detailed control of our actions through the phylogenetically old brainstem structures. For any one area, we can learn the nature of this governance once the destinations and the actions of the layer 5 outputs are known. Often, we know the destinations, but know little or nothing of the actions, specifically for those of the motor axons that have thalamic branches.

It will be important to distinguish the corticofugal axons from layer 5 that send branches to the thalamus from those that do not link to the thalamus for a further relay to the cortex. The distinction is important because the motor actions of the branched axons are subject to immediate monitoring by the cortex, whereas the actions of the unbranched

---

[79] The forward receptive fields discussed in Chapter 13.

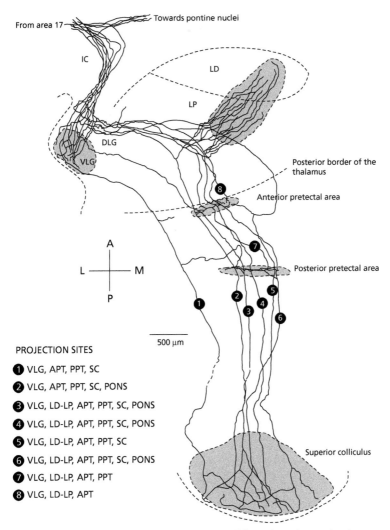

Figure 9.1 Drawing of eight axons that were traced from the primary visual cortex (area 17) of a rat after they had been labelled by an injection of biocytin into cortical layer 5, showing the thalamic branches to the lateral posterior (LP) and lateral dorsal (LD) thalamic nuclei and the long descending branches. The projection sites of these long motor branches are listed at the bottom of the figure and the authors have added that other branches were given off in the internal capsule.

APT, anterior pretectal area; PPT, posterior pretectal area; SC, superior colliculus; VLG, ventral lateral geniculate nucleus.

Reprinted from *Neuroscience*, 66 (2), J. Bourassa and M. Deschênes, Corticothalamic projections from the primary visual cortex in rats: a single fiber study using biocytin as an anterograde tracer, p. 261, Figure 8, http://dx.doi.org/10.1016/0306-4522(95)00009-8, Copyright © 1995 Published by Elsevier Ltd. with permission from Elsevier.

axons will be under no immediate cortical supervision; their roles in ongoing behavioural and cognitive functions are likely to differ.

9.2.1 **A note on the methods that have demonstrated the branched corticofugal axons**

The branch points are not easy to identify nor are the axons easy to trace. For this reason, we need far more information than we currently have. It is important to understand the difficulties of identifying branch points by earlier and current methods. They cannot be identified in sections where large numbers of axons are stained, because a dense population of axons obscures the detail needed for recognizing branch points. One has to rely on methods that select a few axons and then look for evidence of branches on these, tracing the axons through serial sections.

Many of these branching patterns of thalamic afferents were originally demonstrated by the Golgi method, which stains only a small proportion (about 1%) of the nerve cells and their processes (e.g. see Chapter 2, Figure 2.6(a–c)). The Golgi methods have been available since the 1870s and much evidence about branching patterns was recorded more than 100 years ago. Some more recent methods use a marker molecule that labels the axon and generally enters all of its branches (as for Figure 9.1). These markers can be injected into single cells or small groups of cells and their axons can then be traced through serial sections and branches identified. Some branches have been identified by recording from a single cell and stimulating at two sites in order to find action potentials that travel back to the cell and interact with each other.

The Golgi methods and the marker methods have provided most of the currently available evidence regarding cortical afferents to the thalamus. All these methods may fail to show an existing branch. Just one missing or damaged section can cause a branch point to be lost. For any of the available methods, finding a branch is strong evidence that the relevant type of axon has such a branching pattern, but the negative evidence is not proof that the axons do not have branches. It may prove impossible or very difficult to show that all thalamic driver inputs have motor branches. The most telling may be evidence, such as that cited in the next paragraph for corticothalamic driver axons, indicating that more than 90% of a particular group of these axons show such branching.

At present we cannot assert that there are no pure, unbranched driver afferents to the thalamus.[80] However, there is some evidence for a few systems that for a particular class of driver, most of the inputs to the thalamus arise from motor branches. Thus, for first-order relays, Ramón y Cajal's account (1955) of the branching patterns of dorsal root axons entering the spinal cord (see Chapter 2, Figure 2.6(a)) included many sections from a great many different vertebrates as well as the specific statement that he had only seen two ascending axons that lacked local spinal branches. In more recent times, evidence for two systems of higher-order relays has been studied quantitatively (rat somatosensory and cat visual higher-order pathway (Veinante et al. 2000; Guillery et al. 2001, respectively). It was found that more than 90% of the corticothalamic axons having the characteristics of drivers to the thalamus[81] are branches of axons travelling to lower centres.[82] In view of the difficulties of tracing these pathways, it is reasonable to conclude that there are few, if any, cortical driver inputs to these higher-order thalamic nuclei that are not branches of long descending axons. However, we need further comparable quantitative studies of many more pathways before we can claim that essentially all driver afferents to the thalamus are branches of descending axons. And we need evidence concerning the termination and the functions of the descending branches.[83]

---

[80] If any such pure sensory axons were identified in first-order thalamic nuclei it would be important to compare their functional properties and connections with those that have branches, since these would truly represent a pure sensory input in the sense of textbook accounts.

[81] Either an origin from a cortical layer 5 cell or a structurally characteristic terminal in the thalamus, or both.

[82] Superior colliculus or pretectal regions of the brainstem or unidentified brainstem centres.

[83] Some years ago I spent some time in Tokyo at the Riken Institute studying some of K Rockland's sections from monkey brains showing many single axons that could be traced as driver inputs (as defined in this text as coming from layer 5 or having characteristic thalamic terminals, or both), from temporal and parietal cortex going to higher-order thalamic nuclei. Essentially all of these axons had branches that could be traced towards centres in the brainstem. I understand that the sections are still available for study (for further information please contact Dr Rockland: krock@bu.edu).

9.3 **The functions of the branching axons**

The significance of the branching patterns has been mentioned already: the same sequence of action potentials is passed from a parent axon to both daughter branches (Cox et al. 2000; Raastad and Shepherd 2003). This means that the message that the thalamus relays to the cortex will necessarily include information about the instructions that are travelling towards a motor centre: that is, the higher areas of the cortex will receive information about likely upcoming motor events, and their sensory consequences, that are currently being generated by lower cortical areas innervating the phylogenetically old parts of the brain.

We can conclude that essentially all of the messages that reach the cortex from the thalamus come in this dual form: they are about an event in the body or the world and also about a forthcoming action, and, by implication, about a sensory change that will be produced by that action and will, in turn, be likely to generate a further action. Our conscious lives depend on these dual inputs *about how we interact with the world* in the present and the immediate future. It is important to see that this provides a temporal continuity to our conscious lives. Our perceptions produce actions and also depend on actions; these are the sensorimotor contingencies described by O'Regan and Noë (2001).[84]

Whereas all of the messages about the body and the world that reach the cortex depend on the condition of the thalamic gate, the messages that the cortex sends out for our interactions with the world are not

---

[84] Riggs (1976) has described experiments showing that when a visual image is artificially stabilized on the retina by equipment that moves the image accurately to compensate for every ocular movement, perception of the image is lost but the evoked potentials recordable from the visual cortex continue as before the image was stabilized. Riggs suggests that the perceptual loss must occur at a higher cortical level. The evidence described in Chapter 13 would seem to confirm this, but we can surmise that when ocular movements no longer produce the normal sequence of anticipated actions and the anticipated reafferences, the evoked potentials would remain but the perceptual process will be disrupted, rather as it is when reversing lenses are first used (see Chapter 12).

themselves under the control of the gate.[85] The cortex, after receiving a copy of an error, can make a correction to later messages, limiting the error to a brief interval, but it cannot correct the error. Presumably, when corticofugal motor axons that have no thalamic branches make an error, the interval will be significantly longer.

In so far as we treat perceptions as conscious processes that depend on the cerebral cortex (discussed in Chapter 11), the evidence considered here does not demonstrate the standard view that actions generally depend on perceptions. Rather, it shows that actions and perceptions are commonly closely related and represented in a continuous sequence of neural events, each depending on its predecessors. Habitual actions, well-practised actions, and reflex actions often do not depend on the cortex, although the cortex may receive information about such actions, and that information will come through the thalamus as a dual sensorimotor message.

We can compare the standard view, where the incoming messages are passively received from the world and are perceived as images of the world, with the sensorimotor view, where the incoming message comes together with an anticipation of an action and of a new perception produced by that action. The former has us receiving and understanding signals from the real world, whereas the latter has us interacting with an unknown world so that we can use these interactions to learn about a world that matches our anticipations.

Consider the difference between receiving the message 'red patch out there' and receiving the message 'red patch at the expected site'. The former would be a message about the world itself, whereas the latter is a message about our interactions with the world and it may be our anticipation that is an essential part of conscious processing, which is discussed further in Chapter 13.

It is necessary to conclude that the majority of messages relayed to the cortex from the thalamus, not only from the first-order pathways considered earlier, but, more significantly, also from the higher-order

---

[85] That is, as shown in Figure 9.1, they form no synapses in the thalamus.

pathways, come in a dual sensorimotor form to the cortex from the thalamus. If the two parts of the message are ever separated, and if our innate and very strong conviction that they are separate is something more than a culturally rooted belief, then we should look for a separation that occurs in the cortex, probably at a fairly early stage of cortical processing.[86]

---

[86] Notice, however, that if many layer 5 driver outputs from the cortex to thalamus are branches of long descending corticofugal axons, this once more brings action and perception together as a single message at higher cortical levels, suggesting that, no matter what neural mechanisms may locally separate action from perception, there remains a theatre where they are for ever linked. It is possible that some cortical circuits treat the information as about the body and the world and others respond to the information about the anticipated action. So far, investigators have only investigated the first of these possibilities for the sensory pathways.

Chapter 10

The hierarchy of cortical monitors

10.1 Summary

This chapter explores the significance of the dual meaning of the driver inputs to the thalamus in more detail. What happens to these messages when they reach the cortical hierarchies? Currently we know little about how the cortex reacts to the two meanings of the incoming messages. The **efference copies** that reach the cortex may act both in the control of movements, as do efference copies in other parts of the brain, and may also act to generate a conscious anticipation of an action and its sensory consequences. Or it may do both, depending on the circumstances. Where the thalamic relay fails for any reason while the motor branch remains functional, actions may be assigned, as in schizophrenic patients, to external agents. For any one cortical area, we need to understand not only the messages it receives from the thalamus but also the motor instructions it sends out and how these fit into the cortical hierarchy.

10.2 What role do the copies of motor instructions play when they reach the cortex?

We have seen that messages reaching the thalamus for relay to the cortex have a dual meaning: one a sensory message and the other a message about a motor instruction on its way to execution. There is, at present, no reason to believe that on their way to the cortex the two are ever separated, except in neuroscience textbooks and essays on perception.

These dual messages are important if the primary function of the brain is to serve our sensorimotor interactions with the world. However, they make little sense if one regards the primary recipient sensory cortex, by definition, as having specific visual, auditory, or other sensory functions, not equipped also to interpret the meaning of an efference copy. The standard view treats one pathway as carrying sensory information to the cortex, and another pathway and *its* branch as carrying a motor instruction, away from the cortex; that is, the function of the message is defined in terms of the *presumed* capacities of the regions receiving each of the branches.

Where copies of motor instructions are recognized in the standard view at lower levels of the brain, they have been treated as relevant for the control of movements: that is, as 'efference copies' or 'corollary discharges' (Sperry 1950; von Holst and Mittelstaedt 1950; Wolpert and Miall 1996; Wolpert and Flanagan 2001; Wurtz 2008).[87] They have been seen as playing an important role in providing feedback information that is crucial in the production of smooth and accurate movements. If the movement affects a sensor, the expected changes in the sensor's output (the reafference) can be compared with the changes expected from the motor instruction and corrections applied.

Efference copies are present in the spinal cord and the cerebellum and play a significant role in motor circuits generally. They are not a special characteristic of thalamocortical pathways, but in this chapter I am discussing specifically the efference copies that are on the way to the thalamus for relay to the cortex. It is the function of this particular group of efference copies that will be under discussion. I am treating a non-thalamic branch of a known driver input to the thalamus (that is, of a thalamic input that is relayed to the cortex) as having a motor action if it goes to a centre having known motor effects, even where the precise action generated by this axon is not yet clearly defined.[88]

---

[87] Sherman and Guillery (2013) used 'efference copy' in order to exclude activity in multiple sensory branches and I follow this use here.

[88] For example, the *actions* produced by individual branches of retinothalamic or cortico-thalamic driver axons that go to the midbrain and innervate the superior colliculus or pretectum can rarely be identified on the basis of current knowledge. They may produce changes in the lens or the pupil, or movements of the eyes. Each of these actions will change the perceived visual scene. Other inputs to the superior colliculus may, as

An important point about the branches that form the *thalamic* driver afferents from all levels of the hierarchy, concerns their two distinct functional roles: (1) concerning the sensorimotor messages they relay to the cortex, and (2) the information about forthcoming motor actions that they carry.

10.3 **The cortical hierarchy as generator of conscious processes**

We cannot make a clear dogmatic statement about the parts of the brain that play a role in our perceptual and cognitive capacities but it is reasonable to consider the cerebral cortex as a crucial component and to treat other parts of the vertebrate brain as playing no role or an insignificant role. On this basis, the dual messages that the thalamus passes to the cortex are of particular interest for understanding a specific feature of our conscious lives: these messages can provide a mammal with a perceptual experience in relation to its own upcoming actions, providing a distinction between its own actions and the actions of others. That is, they are concurrently not only playing a role in providing a conscious experience about the world, but also in generating our sense of self in the world. The thalamic inputs to the cortex are able to provide a mammal having a sufficiently rich set of cortical areas, with an awareness of its own actions before they occur and consequently of itself *in relation to the world*.[89]

---

shown by Philipp and Hoffmann (2014) (see Chapter 4) produce movements of the arms. The major point is that we know that motor fibres to the midbrain tectum can produce changes that can be anticipated in the cortex because the thalamic branch carries a copy of the mesencephalic action. Learning about the motor actions of the non-thalamic branches, where they have a cortical origin, is likely to prove crucial to a full understanding of the functions of any cortical area with such branching axons.

[89] Since much of the control of movements is taken care of at lower centres it is likely that the cortical role in the control of movements may prove minor, being concerned with fine adjustments or rare errors, and not easily demonstrated once movements have been well learnt and controlled at lower levels. The role in awareness of self-originated movements is perhaps the more significant function particularly in higher cortical areas.

The extent to which there may be non-mammalian organisms or non-cortical parts of mammalian nervous systems that can also generate such a sense of self must be left as

It is possible, indeed likely, that the efference copies that reach the cortex have not given up their role in the control of movements. However, given the capacities of the subcortical regions to control movements (see, e.g. Grillner 2015; Kawai et al. 2015), and the long evolutionary history of them doing just that, it may well be that the need for cortical involvement in the control of ongoing movements is rather limited, especially for movements that have been well learnt.

So far as the branched axons serving as drivers to the thalamus, particularly those going to higher-order thalamic nuclei, it is their role in *cortical* functions that will dominate thoughts about the conscious lives of mammals, and it is their role in *motor actions* that will dominate thoughts about the behavioural capacities of mammals. The reader will by now be aware that we need to learn about the actions of the motor outputs in order to gain useful insights into the functions of the relevant cortical areas.

10.4 The hierarchy of cortical areas monitoring the outputs of lower centres

The messages that are relayed to the cortex by branched thalamic driver inputs must be seen in relation to the corticothalamic connections contributing to the control of the thalamic gate, mentioned in Chapter 8. The transthalamic corticocortical connections differ from the direct corticocortical (Van Essen and Andersen 1990; Van Essen et al. 1992)[90] connections not only because they have non-thalamic motor branches but also because they include the thalamic gate whose actions, as discussed in Chapter 8, depend on whether the thalamic gate is open or

---

speculation beyond the range of this book. I am inclined to treat the mammalian brain as having no such functions at subcortical levels and to leave the question of other centres that may be developed in non-mammalian vertebrates or non-vertebrate animals as entirely open and currently largely unexplored except to a certain extent in birds (e.g. Jelbert et al. 2015; Logan et al. 2016). The reader can be equally arbitrary on the basis of available evidence. I know that investigators who have studied the behaviour of fruit flies closely, find it easy to accept that the flies have conscious experiences.

[90] Although, as indicated earlier, the match of the two hierarchies is only defined for the lowest levels.

closed or has just being opened by a wake-up call. Given the reciprocal topographic arrangements of the cortical layer 6 inputs to the thalamus and to the thalamic reticular nucleus, the well-localized 'searchlight' that can be generated by these reciprocal connections can influence the opening or closing of specific parts of the thalamic gate in relation to limited parts of the body or the world.

It is through these topographically specific connections, not included in Figure 10.1, that each central source of motor outputs has its motor actions monitored by a higher level. We can reasonably expect that the analytical power of the hierarchy and its capacity to react to current events in relation to past experiences will increase significantly as the number of levels in the hierarchy increases (this is discussed further in Chapter 13).

Janušonis (2015, page 2) in a perceptive and generous review of Sherman and Guillery (2013) wrote about this hierarchy:

> [T]he thalamus receives inputs from the actual physical world and its virtual model in the cortex. These inputs map back into the physical world (through their motor collaterals) and into its virtual model (through thalamocortical projections).

This is neat and illuminating but is based on the actual physical world of the standard model. It is useful here to rephrase it in terms of a sensori-motor model: the thalamus first receives inputs from the dual messages carried by drivers to first-order thalamic relays about our sensorimotor interactions with the world and then from the dual messages carried by the cortically generated (learnt?) higher-order corticothalamic drivers about our actions in a model of the world. These thalamic inputs are relayed to higher areas of cortex which can link back to ongoing sensorimotor interactions through their motor outputs and forward to yet higher levels through corticothalamocortical projections (Figure 10.1). At each cortical level the anticipation of the upcoming action can generate a sense of the self in action in a model of the world.

For the discussion in Part IV, it is important to recognize this aspect of cortical organization. Perhaps it can be usefully compared to a large business, where each level monitors the outputs of the next lower level and can generate an alert (a thalamic burst; see Chapter 8) when

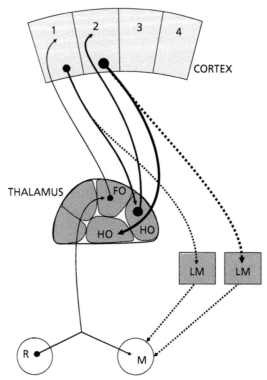

Figure 10.1 This figure illustrates the hierarchy of cortical areas that monitor the motor outputs of lower areas. Only four cortical areas are shown. Each receives thalamic inputs but only the first two of these inputs are illustrated. The thickness of the lines into and out of the thalamus relates to the hierarchical level of the message and thus to its relative 'power' in relation to lower levels. Each cortical area receives messages about activity in lower centres, and each of these messages includes information about upcoming actions, shown as dotted lines for the cortical motor outputs to the muscles. FO, first-order thalamic relay; HO, higher-order thalamic relay; LM, lower motor centre; M, motor cells innervating muscles; R, sensory receptor.

something unexpected occurs, as well as a new motor instruction. This role of the cerebral cortex is discussed further in Part IV.

10.5 Anticipating future actions and identifying self-generated actions

The relationship of cortical activity to a future action at first sight seems surprising: can our brains really look ahead and have information about the future, even though this involves only small fractions of a second? The answer to this is now, I trust, clearly 'yes'; but there is also direct experimental evidence.

In Chapter 6 I discussed the 'head direction cells', a small group of nerve cells that bring messages through the thalamus to the cortex about the position of an animal's head, with each cell responding maximally to one specific direction at any one time. These cells play a role in generating the rat's memory of its environment. Their peak discharges in the thalamus anticipate the optimal head direction by 25–40msec (Taube et al. 1996; Taube 2007). That is, a rat can anticipate the optimum head direction and possibly, associated with this, can also anticipate whatever visual, visceral, or emotional changes are associated with that head direction (see Chapter 6). A comparable situation has been described for several areas of the visual cortex, where most of the cells respond to a particular part of the visual field, called the 'receptive field' of the cell.[91] In several of the visual cortical areas there are cells that change the position of their receptive field shortly before the eyes make a very rapid movement, called a **saccade**. In these cells, the receptive field shifts, shortly before the saccade, to the position that will be the receptive field after the saccade (Colby et al. 1996; Melcher and Colby 2008; Sommer and Wurtz 2008; Hall and Colby 2011); this has been described as a remapped visual field or a forward receptive field.[92]

The clear sense that we have of being the originators of our own actions may well depend on cortical areas that receive advance notice of actions and that can, therefore, be regarded as having a capacity to generate a perception that distinguishes self-generated actions from actions not generated by self. The degree to which any one cortical area can contribute to this sense of the self is not known,

---

[91] The receptive field is the region of the visual field from which changes can act on that cell.

[92] Hall and Colby (2011, page 530) note that in these experiments (for the cortical cell being studied) 'No physical stimulus ever appears in either the old or the new RF'. That is not easy to understand if we are considering a nerve cell that relates only to the *sensory* inputs that reach it from the thalamus or from other cortical areas. If, however, the sensory messages that reach the cortex commonly (or always) also carry a message about an instruction for a forthcoming movement, then the neural mechanisms that can produce an anticipation of a future receptive field are easier to understand.

although there is evidence discussed in Chapter 13 that for any one perceived event, higher cortical areas play a stronger role than lower cortical areas.

The idea that efference copies play a role in allowing a distinction between self-generated actions and other actions is not new. Wolpert, Frith, and others (Wolpert and Flanagan 2001; Teufel et al. 2010; Shergill et al. 2005; Blakemore et al. 2002) have discussed this and Blakemore et al. (1998) have pointed out that we cannot tickle ourselves because we have advance knowledge of our movements. These studies left the specific neural mechanisms unidentified. Recently, Vukadinovic and colleagues (Vukadinovic 2011, 2014, 2015; Vukadinovic and Rosenzweig 2012) have suggested that it is specifically the branching driver inputs to the thalamus, carrying a dual message about a sensory input, together with instructions for a related future action, that can serve to provide the cortex with advance information about self-generated actions, and thus allow self-generated actions to be distinguished from the actions of others. Further, they pointed out that failures of synaptic transmission at the thalamic gate were likely to be responsible for failures by schizophrenic patients to identify actions as their own and therefore to ascribe them to others. They based their conclusions on earlier studies reporting thalamic abnormalities, particularly in the dorsal medial nucleus in schizophrenic patients and on studies of drugs that could act at the thalamic gate and influence the transmission of the efference copy to the cortex while leaving the motor instruction intact. This produces the action without its anticipation, leaving no normally occurring perception that the action is self-generated, and causing patients to assign actions to agents other than the self. The role of a hierarchy of normal such links in our conscious lives are considered further in Part IV.

10.6 A brief note on understanding the messages in the pathways of the brain

Understanding the language of the brain is still far from our capacities. I have written about the possible meanings of **neural messages** in the earlier chapters in terms that need a closer look. Currently, the most successful approaches have involved an ability to relate the temporal patterns of the sensory inputs to a nerve cell to the temporal distribution of the action potentials in the cell's outputs as discussed in Chapter 2 (e.g. Figure 2.3). However, as we saw in Chapter 2, even this is not straightforward: problems of understanding the messages that arrive at a neural centre can be simpler than understanding those that leave the centre, and the further we are from the sensory source or from the motor output, the more difficult it is to know what a message 'means'.

A telling example was described by David Hubel (2005) for studies he and Torsten Wiesel made of neuronal responses in the visual cortex to visual stimuli projected onto a screen. They used stimuli known to produce responses in the retina and the lateral geniculate nucleus. These were circular stimuli with a centre-surround organization (see Chapter 2, Figure 2.3). In the cortex, they found no responses to such stimuli at all. It was not until they were inserting a slide that carried a new stimulus that one cortical cell responded to the edge of the slide as they were inserting it into their projector. Then they were on their way to understanding that cortical cells did not respond to the stimuli that are optimal for retina and lateral geniculate nucleus, but instead respond to moving edges or bars as stimuli. They had been 'asking a question' to which the visual cortex could not respond. Of course, they were in a stronger position than are the many investigators who are looking at neural pathways or cortical areas whose functions are far less clearly defined than are the visual functions of the visual cortex.

If one is looking for a definition of the function of any one neural centre then an account of how it changes the input to produce an output may be as good a definition of the centre's function as one can expect, provided one can understand what the inputs and outputs mean in terms of an animal's sensory, motor, perceptual, cognitive, or behavioural functions, or occasionally several of these.

The answers we get will always depend on the questions we ask. This is clear from David Hubel's account and is clear from Table 6.1 in Chapter 6. For the cortex, this means understanding the messages that arrive from the thalamus in terms of their dual meaning and also understanding the actions produced by layer 5 outputs to the several motor centres introduced in Chapter 4.

We have seen that the connections of the thalamus and cortex with the rest of the brain show that our firm views about the world as something separate from our ongoing interactions with it, are not based on raw data passively received by our senses. The messages that the cortex receives from the thalamus about the world depend on our actions and include information about them. We have to learn early in our own lives to abstract our view of the world from our interactions with the world. As neuroscientists, we still have to learn how the brain processes messages that have more than one meaning, and if we accept that this is still a problem, we have to accept that asking the right question about the functions of a centre or a pathway is much more difficult than our predecessors thought.

Part IV

Higher cortical functions

Chapter 11

Relating the neural connections to actions and perceptions

11.1 **Summary**

So far, I have stressed the neural basis of the interactive view. In this chapter I discuss issues raised by Adrian. To what extent can we read the neural facts in terms of their implications for our conscious experiences? Neural events in the cerebral cortex lead to conscious events, including (but not limited to) perceptions of sensory events as well as perceptions of our own upcoming actions. The neural strength of a conscious event varies, increasing up the cortical hierarchy. Whereas the standard view may seem natural in terms of our own cognitive and behavioural experiences, and there are classical pathways to support the standard view, there are, in addition to the neural reasons considered in previous chapters, many non-neural reasons for questioning this view and taking an interactive view. I summarize these reasons briefly and then look at some of the many novel questions that are raised about the nervous system by the interactive view.

11.2 **Perceptions as cortical events**

In what follows, I will use 'perception' to describe a conscious event and contrast this to sensory events or afferent messages, which need play no role in the conscious life of an organism. For a neural event to be a conscious event in a mammal it must involve the cerebral cortex: all of the evidence from cortical lesions and stimulations leads to that conclusion. Conscious reactions to sensory stimuli are lost after cortical lesions in a highly specific, well-localized pattern and, correspondingly, cortical stimulation of sensory cortex can produce highly specific, localized perceptual responses. In contrast, many responses to activity in sensory or afferent nerves, for example the spinal reflexes, survive cortical lesions even though the normally linked perceptual response is lost.

A further important point about perceptions is that the degree to which we are conscious of an event varies. It is not an all-or-none event. A spinal reflex involves an afferent input that produces a motor output; however, there is no *necessary* perception. A perception of the reflex requires something additional: specifically, a pathway to the cortex *about the reflex*. A dog that scratches at a flea may do so as a reflex act. Sherrington showed that an afferent stimulus imitating a flea crawling along a dog's flank can produce an adequately aimed spinal scratch reflex even when the spinal mechanisms are isolated from the rest of the brain (Sherrington 1906): that is, without the dog's conscious awareness. The actions of a normal dog can change from such a pure reflex act, with the dog not at all concerned about the scratching limb, then to a perceived stimulus as the rhythm of the scratch reflex first changes, then the dog begins to turn its head slowly in the direction of the irritation and finally turns angrily towards the spot and bites at the flea in what looks like a deliberate focused, conscious act. Such action sequences can be treated as acts that gradually cross the borders between an afferent input and a perceived input. Defining these borders in neural terms is impossible on the basis of current knowledge. The important point is to recognize the graded effect. The borders are not sharp or clear for an observer of the dog nor, probably, would they be for the dog itself or for a physiologist studying the dog. We still need to define the extra components that enter into the actions as they shift from purely reflex to apparently conscious ones, but there is nothing, in principle, to prevent an account in purely neural terms.

The following account will include this graded feature, and I will suggest that the level of the cortical hierarchy involved may be the most relevant: higher cortical areas, HO in Figure 10.1 and higher levels not shown in the figure, are likely to involve sequentially higher levels of consciousness, and this can be expressed from the point of view of the neuroscience as more nerve cells being involved in the act at higher levels of the cortical hierarchy (see Section 13.3 in Chapter 13).[93]

I can walk into a room and be conscious of several features in the room and be quite unaware of others even though I can see them and may even touch them. The discussion that follows is about perception as involving conscious awareness and the capacity, for people, to report that awareness. Without recognition of a distinction between reception of incoming signals (sensation here) and a conscious experience of the world, accounts of the relationships between action and perception easily become more mysterious or confusing than they need be.

There is one other feature about perception that has a long history in philosophic discussions. Although it may prove possible to match specific perceptual experiences to well-defined neural events, and to show that one perceptual experience differs from another in terms of identifiable neural variables, the perception itself is a private event unknowable to others. Perceptual experiences are first-person experiences, sometimes referred to as qualia, and a significant amount of artistic and literary talent has been devoted to efforts at communicating the nature or the contents of these first-person experiences. They may at first appear to be simple, such as the colour of an apple, or they may be complex, such as the pain produced by an injury, a failure at an important task, a romantic attachment, or the death of a friend. These have challenged the descriptive skills of many great artists and authors. The literary accounts can help us to understand these private experiences, even when they are quite simple, but they do not provide an objective (reductionist) account of the experiences themselves. From our point of view here, it may be possible to match them to neural events (psychophysical parallelism), but in themselves they can only be known to the perceiver.

---

[93] It is not possible completely to exclude subcortical centres from being relevant for the generation of conscious experiences, but if they are, then it is likely to be extremely weak relative to even the lowest levels of the cortical hierarchy.

11.3 **The origins and limitations of the standard perception-to-action view**

It is important first to look briefly at the possible origins of the standard view before we consider how our account of the thalamocortical connections relates to earlier challenges to the model. I say possible origins, because, as I suggested in Part I, the standard view seems so natural to a Western mind, that it may seem perverse and unnecessary to look for an origin. We look at the world, we hear it, touch it, or smell it, and then we act. We see ourselves as captains of a battleship who watch the weather and the enemy and who issue orders accordingly. The world is out there and we have to find out about it before we can act. The standard view seems entirely reasonable. We have seen that the sensory pathways to the cortex and the motor pathways from the cortex fit this view so that it may seem foolish to ask about how such a view could arise or, for that matter, to challenge it.

Although it is easy to make a case that the neural connections can support the standard model and to think of it as strongly based in neuroscience, once we recognize that from their first entry into the spinal cord or the brain, the afferent pathways have branches that link to the motor systems, we face the resulting dual nature of the messages reaching the cortex through the thalamus and this inevitably raises questions about the standard view.

The many motor outputs, other than those from the classical major motor areas, going to the lower centres (some are shown in Figure 4.1 in Chapter 4) further suggest that something is missing from the standard view. Most strikingly, the observation that copies of many of these cortical outputs to lower motor centres are relayed through the thalamus back to higher cortical areas, establishing a hierarchy of *cortical monitors of motor actions*, raises new issues, specifically about the neural connections that do not fit the standard view. It is the neural connections that are my major focus. Understanding these connections in relation to the standard view not only forces us to question the standard view, but it also raises important new questions about the structure and function of the existing connections, questions that will need answers from new experimental studies of brains. We will need to understand the neural pathways that form the hierarchy of monitors. There are many unknowns concerning these connections and their functional roles. Finding answers should prove wonderfully challenging and productive.

The important point that will be added here is that the standard view has also been challenged by many on the basis of observations that generally do not relate to the neural connections. The following brief summary shows that arguments based on the neural connections do not stand alone.[94]

Merleau-Ponty (1958), writing as a philosopher with an interest in art, recognized the extent to which perceptions were not passively received by us. He wrote that an object presented '... to the gaze or the touch arouses a certain motor intention which aims not at the movements of one's own body, but at the thing itself from which they are, as it were, suspended'. This curious phrase makes sense more readily for a tactile exploration in the dark, where the movements of the fingers 'aim' at the object that needs to be perceived; but once that is recognized it can also be seen to be relevant to the movements of the eyes that are exploring a new object.

Gibson (1986) wrote, 'we are told that vision depends on the eye, which is connected to the brain ... I shall suggest that natural vision depends on the eyes in the head on a body supported by the ground ... we look around, walk up to something interesting and move around it so as to see it from all sides, and go from one vista to another. That is natural vision'. He argued that information about the self is closely linked to and inseparable from information about the environment and developed a theory of 'affordances' that provided for our particular interactions with components of the environment.

Churchland et al. (1994) have argued that it is a common error to think that 'the business of the visual system is to create a detailed replica of the visual world' in the brain. Perceptions cannot, they suggest,

---

[94] The contributions made to an interactive view of perception by Hulme and Helmholtz are discussed in Chapter 1.

be treated as 'pure' representations of the world received by a passive organism. Their alternative is 'interactive vision'. They point out that from an evolutionary view vision is there to provide improved motor control. They stress the important role of visual learning for identifying items in the world. They use a number of illustrations, including classical illusions and strange sensory outcomes of experimental situations. They mention the large number of cortical areas that have motor outputs to the basal ganglia or the superior colliculus, but their main point deals with the behavioural and cognitive issues, not with the details of the relevant neural connections.

O'Regan and Noë (2001) stressed the extent to which our actions determine the way in which we see the world. They pointed out that the rules that govern perception differ for each sensory modality, describing these rules as based on 'sensorimotor contingencies', where the perceptual inputs and the actions produced by the outputs depend on each other. They treat seeing as a skilful activity that depends on the extent to which one has mastered the rules relevant for that sensory pathway. This skill, which, for vision, includes the ability to control the movements and the functions of the eye, presumably having to be learnt.

Hurley (1998, 2001) challenged the validity of what she calls the 'sandwich model of the mind', with the mind forming the filling between the way in and the way out. She describes situations for which the sandwich model does not account and proposes that perception and action are really not peripheral and separate from one another, basing her answer on neural network and dynamical systems approaches, on observations of sensory prostheses (considered in Chapter 12) and also on theories concerning feedback or circular causal flows.

Mossio and Taraborelli (2008) have questioned the extent to which action and perception are related to each other. Do our perceptions depend specifically on our actions (reafferences)? Or do they also depend on changes in our environment (exafferences)? They treat the problem in terms of a 'hypothesis [that] questions the traditional distinction between perception and action as independent cognitive domains' (page 1325). In those terms, we can understand that our perceptions can be produced in the absence of our own actions, as when we are listening to Bach or experiencing a thunderstorm, but once we know that our perceptions depend on branching axons that carry instructions for actions, it becomes necessary to reject the hypothesis.

There is an extensive literature (e.g. Varela et al. 1993; Varela 1996; Thompson and Varela 2001; Noë 2004; Clark 1998, 2008) on interactive or embodied perception, which sees the extent to which our perceptions (and our conscious processes in general) cannot be limited to the brain, but must involve the whole body interacting with the world, that is, we depend on neural records of such interactions, in which the body, the brain, and the world play a continuous role.

The reader who has reached this point will know that much of what follows will also be a clear 'no' in answer to Hurley's question about the separate nature of perception and action. The several points made by the above-mentioned authors and others must now be related to the anatomical and functional relationships summarized in the earlier parts of this book. These include the dual messages that the thalamus relays to the cortex about sensory information and anticipated actions. They also include the functions of the thalamic gate, the many cortical layer 5 inputs to lower motor centres that contribute to our actions, and it includes the transthalamic corticocortical links that arise as branches from the cortical layer 5 motor outputs. These generate the hierarchy of cortical areas where higher levels monitor both the sensory inputs and the motor outputs of lower levels, relating perception and action to each other at all levels.

In this system, the neural pathways for perception and action are closely and continuously related to each other over time and continuously link back to the body's actions. The messages that the thalamus passes to the cortex are about our upcoming actions in relation to our present and also to our upcoming perceptions. Further, they are about the continuity of these events. They are about our anticipation of the appearance of the world once an anticipated action is completed. Because we have such information about our forthcoming actions, we can anticipate our perceptions with a high probability of getting it right. The occasions when we don't get it right are important. This is when higher cortical circuits are called into action by the thalamic generation of a wake-up call (see Chapter 8) stimulating a search for the origin of the discrepancy[95] so that appropriate higher cortical outputs can be generated for needed corrections. The thalamic gate together with the higher-order cortical motor outputs provides higher cortical areas with the needed executive powers. When we do get it right, this relationship serves to give us, ourselves, a continuous presence in the world over time. When we don't get it right, that continuity can be strangely disrupted, our perceptions are confused or absent, and our actions may appear to be those of an external agent, as illustrated by the experiences of a schizophrenic patient.

11.4 Can further knowledge of the neural connections help us to understand how actions relate to perceptions?

We have seen that dual messages are relayed from the thalamus to the cortex, but we have no information about how they are processed in the cortex. There may be separate pathways for treating the sensory and the motor components, one proceeding from cortical layer 4 to layers 2 and 3 then processed for sensory content by the direct corticocortical connections, the other component proceeding to layer 5 to be issued as a motor instruction to lower centres.

For any one cortical area we need to learn the actions of the layer 5 outputs: what are the specific motor instructions and how do they relate to the thalamic inputs to that area? Or, for a sensory pathway, to the perceptions? Some of the motor outputs are likely to lack branches to the thalamus, but others will send branches to the thalamus for further cortical monitoring in higher areas. This can provide an opportunity to compare actions that involve the thalamic gate with actions that do not.

An important issue that has not been raised yet concerns the extent to which the involvement of a cortical area will vary as a new sensorimotor relationship is being learnt. For any one cortical area, how important a role in the control of movements does the layer 5 output of that area play? To what extent does cortical involvement of a well-learnt response

---

[95] It may well be that there is no perception until the discrepancy is resolved, but that is something that needs to be checked experimentally.

differ depending on the hierarchical level of the cortical area? To what extent are well-learnt actions executed by the phylogenetically old brain, with minimal involvement of any cortical areas?

Perhaps most challenging, we need to learn how the neural links are formed that produce (teach us) the strong sense of a permanent world out there (perceived colours do not depend on the illumination, shapes do not depend on the angle of view). This is a question about infants learning the constancies of the world; it is also a question about adults adapting to new sensory experiences discussed in Chapter 12, but for students of the nervous system it is about the properties of neural circuits.

Chapter 12

Interacting with the world

12.1 **Summary**

In this chapter, the extent to which actions and perceptions depend on each other is explored particularly for the visual system. Viewing the world through a mirror or a lens that displaces or inverts images provides examples of our ability to learn new sensorimotor consistencies. The use of sensory prostheses that replace one sensory modality with another, for example, visual by tactile stimuli or vestibular by tactile stimuli, provides examples of the capacity of our brains to learn about new sensorimotor relationships, often with surprising rapidity, even in an adult.

12.2 **Helmholtz**

In modern times, an early suggestion that perceptions depend on actions was presented by Helmholtz (Warren and Warren 1968). One of his illustrations of the problem referred to the way we sometimes look at the world with our head tilted to one side, upside down from under our arm, or if we are young children, from between our knees. This upside-down view of the world looks strange and fascinating even though the world is the right way up and it is only the head that is upside down. But something about the view is changed, and Helmholtz stressed that, especially for appreciating the relationships of colours in a scene, in the novel view, with the head tilted or upside down the colours relate less closely to the viewed objects and can be seen simply as perceptions in their own right.[96]

---

[96] See earlier note on the confusing use of 'sensations'. Here the important distinction is between a *perceived* object that we have learnt to treat as not changing colour and the interpretation of the neural messages about colour that are generated when the retina is upside down, changing in unexpected ways as we move our eyes or the object moves.

Another, neurally based reason for the queerness of the view was suggested by Bompas and O'Regan (2006; and see Mossio and Taraborelli 2008) who drew attention to the fact that our views of any scene depend on the distribution of the different classes of visual receptors for specific wavelengths over the surface of the retina. These receptor types are not each distributed homogenously nor are their distributions symmetrical about the horizontal meridian. Thus, the sequence of stimuli relevant for colour perception produced by eye movements with a normal head position are not matched by the sequence of stimuli with the head upside down, and the perceptual result of the unusual upside-down sequence will be unexpected. The strangeness of the colours perceived are due to the unexpected sequences of the information about the wavelengths received by the retina, which differ from normal in the upside-down head position.[97] If the retinal receptors were evenly distributed over the surface of the retina there would be nothing odd about the upside-down view, nor would there be if the distribution of the receptors were symmetrical about the horizontal meridian of the retina. We have got used to viewing the world with our heads in a more or less upright position and in some way we *have learnt* to compensate for the changes in inputs that accompany every ocular movement. We can use the tilt or the inversion of the head to challenge the usual, expected view of the world when we want to try to understand a more detailed view of the colour structure of a scene or a painting, or, when as children, we simply want to be puzzled by the new and strange view of the world.

The conclusion is that, as we move our eyes (normally three to four times per second or so) and view any one object with different parts of our retinas,[98] the *perceived* colour of the object does not change. Although when, in normal vision, we look at an apple, the combinations of stimulated wavelength receptors changes as we move our eyes or the apple, we generally see the same red apple. We appear to have

---

[97] The details of how we perceive colours are beyond the scope and the immediate needs of this book. Interesting treatments can be found in Land (1974), Zeki (1993), Hilbert (2005), Reeves et al. (2008), and Foster (2011).

[98] Parts that cannot even be matched in binocular terms for their receptor distributions except possibly for one viewing distance.

adapted to these changes, and have probably learnt to adapt to them from our earliest experiences of seeing the world. We have learnt to assign perceived properties to 'objects', that is, to bits of the perceived world whose continuity plays an important role in our interactions with the world.[99] These are items that appear repeatedly, and information about them needs to be stored in memory. We know how to interpret the colour of objects as independent of our ocular movements. We early learnt to discount the changing patterns of retinal stimulation so that we could deal with the world in terms of objects that could be treated as stable items. Exactly how we do this in neural terms is not clear, but the fact that we do it, that we learn to identify objects in terms of their apparently constant colour, shape, and size cannot be questioned. It is a challenge for neuroscientists. We know that a round, green plate maintains its apparent size, colour, and shape no matter how we view it. There can be changes in the way the plate is illuminated, it can be close or on the other side of the room, it can appear as a circle or an ellipse, it always appears to be the same plate having the same colour and shape. Where investigators have found counter-examples, of items that do not maintain a constant appearance, these are often described as visual illusions (e.g. Churchland et al. 1994).

Helmholtz also considered the problems of wearing spectacles with prisms that deflect the view of the world to one side. He described the problems that arise when the wearer needs to match the movements of the hand to the new visual inputs from the hand, and he adds that a similar problem arises when a young man first learns to shave in a mirror or a young investigator first tries to manipulate objects under the (inverting) lenses of a microscope.[100] At first, the problems of wearing prism spectacles are serious, and controlling the movements of the

---

[99] I saw a grandmother on the bus this morning with a very young child, and a box full of wooden blocks varying in colour and shape. The grandmother was picking out blocks, was holding each block still, and was naming the colours. She had the impression that she was teaching him the names of colours, but I had the clear sense that he was learning to see shapes and colours and to learn the constancies. He was not listening, he was holding another block looking very closely at this as he moved it about, ignoring her completely.

[100] Helmholtz published an important early study of nerve cells as seen under a microscope. He used the nerve cells of the leech, readily available then to any young doctor.

hand as well as making the appropriate movements of the eyes in the novel visual field can be frustrating. The perceptual results are confusing and inconsistent. There is a loss of the wearer's normal perceptual skills. However, the movements of hands and eyes gradually become natural, and when the spectacles are finally removed the movements once more become difficult, although the disruption is less severe than that produced by the first use of the spectacles, and normal vision returns. The new relationships are not learnt as well or as thoroughly as those that were first learnt during infancy.

Helmholtz thus clearly recognized and demonstrated the importance of movements in the generation of perceptions. Helmholtz recognized the strong conviction that often stands between our knowledge of the nervous system as a recipient of information about a real world and our ability to interpret perceptual experiences on the basis of that knowledge. That played a significant role in his study of the retina. When we recognize the dual nature of driver inputs to the thalamus, about the world and about upcoming actions, we can begin to understand, in terms of neural connections and functions, what Helmholtz was claiming about actions and perceptions.

Westheimer (2008) wrote that Helmholtz 'continued to assert that our knowledge of the real world is derived by using our motor system as an exploring organ to *deduce invariances by trial and error*. Generating a movement is the activity of the "I" which keeps track of the instructions' (page 647, italics added). It does not matter whether it is our fingers exploring the world in order to identify objects on a table in the dark or our eyes, comparably, exploring the world in order to arrive at an 'unconscious conclusion'[101] about what is out there in the world. Somehow, we (ourselves) this 'I', *is keeping track* of the instructions that move the fingers or the eyes. We, that is, those of us who know about the branching inputs to the thalamus, can understand the neural basis of

---

[101] It is important to stress again that the process of arriving at the conclusion is unconscious, whereas the conclusion itself, the learnt outcome of the sensorimotor interaction is necessarily a conscious act. These conclusions about the structure of the world can perhaps be regarded as the original appearance in evolution, or in an infant's life, of a sense of self and of conscious processes.

this capacity to keep track of the instructions. It is a necessary outcome of these branching axons that bring information about the instructions for forthcoming actions to the thalamus for relay to the cortex. Westheimer added a useful parenthetical explanation of this 'I' keeping track of instructions.[102] He wrote (pages 647–8), 'A modern term is efference copy, meaning the record that is maintained of the outgoing or **efferent** signals from the central nervous system to the muscles'.[103]

Consider Figure 10.1 in Chapter 10, showing a cortical output going to a lower motor centre and a branch going to the thalamus for relay to a second, higher cortical area. This second cortical area will now, on the basis of learning from earlier activity in this particular pathway, have information about the perceptual change that will be produced, after an expected delay, by an anticipated action of the relevant lower motor centre. That is, neural machinery in the cortex is available for learning to anticipate perceptual changes that will be produced by specific actions of the organism itself. In Westheimer's words, these circuits can serve: 'to deduce invariances' by 'keeping track of the instructions' for movements. An issue arises about the extent to which the neural circuits can serve to respond to changes that are in conflict with a learnt view of a constant world. When, for instance, the anticipation fails, as it does when the view of the world is first displaced by a prism or in a mirror, the function of the cortical circuitry is challenged. A thalamic wake-up call may alert the system to the unexpected events, but that will only be the first stage of a long process of necessary re-learning,

---

[102] When in subcortical centres the sensory outcome of an action, the reafference, matches expectations, no corrections need to be initiated. The same may be true when a cortical area is participating in an action. However, we need to think of another outcome of such a cortical activity, which, once the instruction for action matches the expected reafference, may allow for the identification of the self that is producing this action, an identification that can perhaps be regarded as one of the simplest conscious acts.

[103] I quote it here to stress that Westheimer is writing about the same connectivity patterns that Murray Sherman and I described for the inputs to the thalamus for relay to cortex (Sherman and Guillery 2013), except that the cortical locus of the message, and thus its role in conscious explorations is not included by Westheimer. The role of efference copies in defining what here is the 'I' and what I have discussed as the 'self' is treated in Chapter 14.

matching the expected sensory changes (the reafferences) to those actually received by appropriate newly adjusted motor outputs from the cortical hierarchy.

If we ask what an unconscious conclusion about the world actually is, the best answer may be that it is a relationship that has to be learnt so that our conscious perceptions of the world can hold constant those features of the world that we have learnt to regard as constant on the basis of our interactions with the world. The key point may well be that we form the conclusions unconsciously, but that the conclusions themselves are necessarily conscious. They provide our conscious views of the world, and provide an essential entry into our conscious lives.

For any input, even when one or another feature, apparent colour, shape, or size, is varying, we can identify the constancies. The infant may have to learn that particular movements of the eyes produce particular perceptual changes; that objects don't change colour as we move our eyes or our heads, or as the objects move, even though a different combination of photoreceptors is being stimulated by that object[104] and the light reflected from the object has changed. Adults who first start to wear spectacles with prisms, must *learn* that movements relate to perceptions in new ways. It is reasonable to treat this learning as similar to an infant's learning at first encounters with the world, although, like so much learning, it may be easier earlier even though the adult may have to establish fewer new constancies. Our perceptual experiences are all based on our *interactions* with what we *learn to regard* as a stable world. We learn to understand how perceptual experiences relate to features that we need to treat as constant in order to interact successfully with that world. It is not that we should regard the world as inconstant, the inconstancies are often, but not invariably, produced by our movements and the properties of our receptors. We have to recognize that the constancies we ascribe to the world are based on our interactions with the world, are learnt and are useful to us.

---

[104] There is good evidence that infants have to do a great deal of learning as regards their visual capacities and their movements (e.g. Dobson and Teller 1978; Bronson 1990).

The challenge for the neuroscientist is to explore neural activity as new constancies are being established, first in the infant and, perhaps more accessible to experimental approaches, later in the adult, when, for example, reversing lenses are first worn or new perceptual relationships are established. Perhaps experimental animals can be trained to interact with mirror images, but a more promising experimental situation that has already been developed may prove to be the capacity of ferrets to relearn directional hearing after one ear has been reversibly blocked (Kacelnik et al. 2006; Nodal et al. 2010).

12.3 Some other views on action and perception

Merleau-Ponty writing in 1945 also considered the puzzling changes experienced by people who wear inverting lenses as described by Stratton (1896–1899, cited by Merleau-Ponty (1958)). Merleau-Ponty provided a clear account of the slow adaptation of the wearer to the reversing lenses over several days, during which parts of the world appeared to be the right way up while others seemed inverted. This period of confusion lasted up to about a week, during which the image reaching consciousness varied, presumably as neural activity at first failed and finally succeeded in matching the instructions for movements of eyes, hands, or fingers to produce the expected reafference, and then the problem of the inverted world appeared to be 'cured'; the world could once more be consciously perceived. Then, when the glasses were removed, there was another brief inversion of the world before the subject returned to a normal visual experience. It is difficult to explain these changes without recognizing that a subject's visual perceptions of the world are guided primarily by the movements of the eyes, which, if they are to move from one point in the world to another must follow the rules imposed by the reversing lens in the early parts of the experiments and by their absence at the end of the experiment when the lenses are removed. New patterns of movements of the body, particularly also of the hands and fingers, are necessary to adapt to the inverted image. And the adaptation must be regarded as a learnt response that serves to maintain invariances that are of practical importance in a world created by our 'unconscious conclusions'.

More recently, many other authors, already mentioned in Chapter 11, have considered comparable problems of sensorimotor relationships, including a number of visual illusions that can be interpreted in a way that is either puzzling on the basis of the facts known about the image itself or that is inconstant, varying from one 'unconscious conclusion' to another. These studies were generally not based on specific neural connections and preceded observations showing that efference copies pass to the cortical hierarchies concerned with perceptual processing.

Although today it can be seen that the thalamocortical connections are relevant to the observations reported, the precise links between the phenomena reported and the actual neural connections still remain to be defined.

An additional, fascinating, and deeply puzzling group of problems about perception considered by several authors concerns the perceptual capacities produced by prostheses that replace one sensory modality with another, for example, replacing vision by touch.

12.4 **The evidence from sensory prostheses**

Some of the most striking examples relating perceptions of the world to actions come from visual prostheses, such as that developed some years ago by Bach-y-Rita at Madison, Wisconsin (2004; and see Danilov et al. 2007).[105] This is basically a digital camera attached to the head of a blind person, or a camera that can be manually moved so that the environment can be scanned by movements controlled by the subject. A digital version of the camera image is then delivered as tactile stimuli, originally to an area of the skin of the subject's back, but in later trials to the tongue, which is a far more richly innervated sensory surface. The subject needs to control relatively few movements of the camera before the tactile stimuli begin to be interpreted as messages about

---

[105] See also the video clip at: https://vimeo.com/59755393

'visual' images delivered by the camera to the skin or the back or the tongue. However, if the subject is not in control of moving the camera then there are no such 'visual' images, no matter how much the camera is moved. The perception depends on the subject's actions. A striking example is of a subject who has become used to manipulating the camera being shown a large object that rapidly approaches the camera, that is, it increases in size. The subject will move back, away from the apparent oncoming object even when the relevant tactile stimuli are being delivered to the skin of the back.

In other trials, Bach-y-Rita used a prosthesis that could replicate the stimuli about head movements normally coming from the vestibular apparatus of the inner ear and this, again, transmitted the relevant stimuli to the tongue. In patients who have lost the function of the vestibular sensory receptors, movements become extremely difficult or impossible because we rely on the vestibular inputs to keep our bodies under control as we walk and also to keep our eyes moving to compensate for head movements.[106] Our stable visual image of the world depends significantly on the **vestibulo-ocular reflex**, a reflex that automatically adjusts eye movements to compensate for head movements, so that the world still looks stable when we move our heads. In addition, many other vestibular reflexes play a role in helping us to keep our bodies stable as we move about. We do not consciously have to deal with the action of gravity on our bodies, and to a significant extent we know where our bodies are and how they are moving through the world, in part because of the messages generated by the vestibular organs in the head.[107]

The effect of such a vestibular prosthetic device for patients who had lost vestibular functions was dramatic. Patients who had previously been bed-bound were able to move about freely: there are videos of Bach-y-Rita dancing with a patient and also of a patient riding a

---

[106] For other types of vestibular prostheses and for the relevance of eye movements see, for example, Dai et al. (2011) and Merfeld et al. (2006).

[107] Also, the joint and muscle receptors of the limbs and the trunk are relevant, but are not sufficient for orienting the body or the eyes as we change our position in space.

bicycle. One of his assistants once came to me many years ago, before I had thought much about these problems and asked me whether, based on my knowledge of neuroanatomy, I could identify the pathway that messages might take to travel from the tongue to the vestibular nuclei of the brainstem so that the appropriate vestibular messages could once more be sent to the muscles. I did not have an answer, but at the time I thought it was the wrong question. Bach-y-Rita's answer would have been: 'Just give the brain information and it will figure it out'.[108]

That answer is mysterious in terms of what we currently know about brains and nerve cells but it provides a hint of what we should be looking for: neural mechanisms that can recognize regularities in the world as the sensory inputs vary with the movements of the relevant body parts.[109] These regularities must be learnt *and also committed to a memory store*.[110] That is essentially what our knowledge of the world is based on (until we learn to speak, read, and watch moving images on screens).[111]

Clark (1998), writing about the vestibulo-ocular reflex, suggested that if the computations underlying the vestibulo-ocular reflex of a subject

---

[108] http://tcnl.bme.wisc.edu/laboratory/founder

[109] O'Regan and Noë's sensorimotor contingencies.

[110] Strangely, the relatively normal ability to move about lasts for some time after the vestibular prosthesis is disconnected.

[111] Bach-y-Rita's comment that the brain will figure it out is, of course, a neuroscientist's shorthand for saying that the patients will figure it out by using their brains. The problem for the neuroscientist remains: how is it done? The role of the eyes in providing a stable image of the world and thus a stable image of how the patient relates to the world may provide a useful clue to what is going on. The vestibulo-ocular reflex plays a large role in ensuring that a person's eye movements can be stably related to the head movements. The information about head position transmitted to the tongue provides the missing information about head movements for the patients. What the patient is 'figuring out' is that the tactile stimuli to the tongue relate to the limited information that is still available about head movements through receptors in the muscles and joints of the neck, and also relate to the changing visual images on the retina that are produced by these movements of the head. The patient's capacity to integrate these several inputs so rapidly is astounding, and since the final (learnt) solution, apparently so readily and rapidly arrived at, involves several body parts not merely the brain, it may be more accurate to say that the patient figured it out than to assign this all to the brain! To my knowledge no one has actually worked out the computations that would be needed to answer the assistant's question as to how this is done.

were assigned to some external device, then the relationship between the subject's conscious experiences and the workings of the reflex might be quite normal. This may well be a key to understanding the effects of the vestibular prosthesis. If the information provided by tactile images that represent head movements when fed to the tongue suffice to allow stable fixation of the eyes on a part of the visual scene, that may be enough to provide a stable base for the patient's activities. However, the neural pathways concerned in this surprising recovery of functions remain unexplored.[112]

These several observations about reversed views and novel ways of perceiving the world with prostheses help to show that our experiences of the world depend on learning how our interactions with the world relate to our conscious conclusions about the world that we arrive at unconsciously. This is a key to much that has been written to question or challenge the standard view of perception. Indeed, it can reasonably be argued that the *only knowledge* of the world that we can rationally claim depends on what we learn about the consistent features revealed by our sensorimotor interactions. The idea of 'embodied perception', perception that is dependent on the movements of our bodies and limited by the functional capacities of our bodies,[113] has played a significant role in the modern challenge to the currently dominant neuroscientific view of perception. We need to recognize that our view of the real world 'out there' is not primary data that we receive as information about the world. The primary data for us as biological organisms (not as sophisticated adults with a well-established view of the three-dimensional world that provides our environment) arrive as messages that provide information about changes in the world or the body that come *together with* information about our own upcoming related actions and with

---

[112] There has to be a rapid learning process that links the signals to the tongue to the instructions for head movements and in turn links these to the eye movements. Whether these are events in the parietal cortex or in the midbrain (see Chapter 4) is currently unknown but those are the places one might start to hunt for a neural basis of the vestibular sense provided by the prosthesis.

[113] But including aids such a spectacles or a blind person's stick, microscopes, and so forth (see later discussion).

information about the resulting changes in perceptions that will be produced by those actions (the reafferences).[114] This is all the information that we have in terms of a private, first-person experience before we interpret it (arrive at an unconscious conclusion about it) as being about the three-dimensional world including ourselves.

Information about the sensorimotor interactions that produce successful (or failed) outcomes in terms of our interpretations and interactions with the world, all passes to the thalamus and cortex, and our views of the world are built up from, learnt from, the constancies that we can abstract (conclude) from the raw sensorimotor data.

MacKay (1982, page 285) wrote:

> First among our data, both logically and ontologically, are the facts of our conscious experience.

If we recognize these data as the basis from which we create our view of the world and of ourselves, then it makes no sense to ask for an explanation of these data (that is, of mental events, Hume's 'ideas') on the basis of the world that we have created from them. That would be putting the cart before the horse, explaining the given data by the conclusions they lead us to; as Chalmers and Jackson (2001, page 316) pointed out: 'if the mental is fundamental, it cannot be reductively explained'.

---

[114] It is important to recognize that the reafference brings information to the cortex that is a part of our perceptual experiences. The reafference can contribute to the accuracy of the movement and it can also play a role in our ongoing perceptions related to the movement. These, as indicated earlier, can be considered as two distinct functions.

Chapter 13

The role of the thalamocortical hierarchy

13.1 Summary

This chapter presents evidence that at each level of the thalamocortical hierarchy the strength of our conscious perceptions increases. Conscious processes are not all-or-none effects, they are graded. Four factors may be particularly relevant for understanding the neural production of conscious experiences: (1) the actions of the thalamic gate; (2) the neural activity that anticipates an organism's actions; (3) the activity of the hierarchy of cortical monitors; and particularly (4) the motor actions produced by the outputs of the cortical monitors and acting on the phylogenetically old parts of the brain: these serve to keep actions in accord with anticipations.

13.2 What is special about the mammalian thalamus and cortex?

The view that the efference copies reaching the cortex provide a conscious anticipation of events, specifically of the organism's own actions, is supported by the cognitive losses reported by schizophrenic patients when, for example, they assign their own actions to others. This evidence when combined with evidence, summarized earlier (see Chapter 10), about cell losses in the dorsal medial thalamic nucleus of patients, and of drugs that can block the transthalamic corticocortical link and produce these symptoms, provide a start at an answer about the neural basis of the cognitive capacities of the mammalian cortex. However, it is only a start. We saw in Chapters 9 and 10, that any subcortical efference copy can generate a neural anticipation of an action and of the related reafference. If we ask what the mammalian cortex provides that our non-mammalian, vertebrate ancestors lacked, then a key function may be the sequence of inputs that are passed up the transthalamic hierarchy of cortical areas, with each level not only having an opportunity to monitor lower *cortical* levels, but also able to modify the relevant actions in the phylogenetically older, subcortical parts of the brain. From the evidence available, it seems that even the earliest mammals with the simplest cortical development had their major sensory inputs represented by more than a single cortical area (Kaas 2011). That is, in the early appearance of a mammalian cortex, a hierarchical organization may have been as important as a characteristic laminar structure.

The assumption, that it is specifically the cortex that plays a crucial role in the generation of conscious experiences, must, as indicated earlier, be based heavily on clinical and experimental evidence about the effects of localized cortical lesions or stimulation. However, a further question needs to be asked: what distinguishes sensory events that can reasonably be regarded as perceptions, that is, events of which an organism is aware, from sensory events of which the organism appears not to be aware? And if there is a clear distinction, is it graded or clearly discontinuous, either conscious or not? Two groups of experiments of the visual pathways have suggested that the necessary functional organization is not to be found in any one area of cortex but may be available in most, possibly all cortical areas, from primary receiving areas, to the highest, and that it is graded, increasing in strength for higher levels of the hierarchy.

One group of these experiments has shown that when monkeys steadily view two different images, one through each eye, the images for the monkeys, as for human observers, alternate perceptually (Leopold and Logothetis 1996; Logothetis 1998; Keliris et al. 2010). The monkey can be trained to report which image it is seeing. About 20% of the cells in early visual areas (V1 and V2) respond in accordance with the image that the monkey reports as perceived, whereas in higher visual areas these numbers are higher (V4 and MT: 38% and 43% respectively).[115]

---

[115] In all areas, the majority of cells continue to respond to their preferred stimulus even when perceptually it is dominated by the competition. Areas V1 and V2 were not distinguished because the recordings came from the region close to the border between these two areas.

In another relevant group of experiments, Melcher and Colby (2008) studied the forward receptive fields described in Section 10.4 in Chapter 10 and showed that there is an increase of the cells with forward receptive fields, with very few such cells in V1 (2%), but more in higher areas: V2 (11%), V3 (35%), and V3A (52%).[116]

Thus, there are two important variables, each of which may be related to the salience of the conscious experience. One is the number of cells that are firing and the other is the hierarchical level of the cortical area. The two appear to be closely related, but the relationship may well depend on other, unknown factors as well.

So far, we have no answer to the key question as to how the cortex does this job of generating first-person experiences. On the basis of current knowledge, it makes no sense to be looking anywhere other than cortex for the crucial components, and the inputs must be drivers[117] in order to generate conscious processes *about* the body and the world in the cortex. We need to ask not only how neural activity that merely responds to a sensory input from the body or the world differs from activity about events that are perceived, but also learn how the number of cells that relate to perception is increased up the cortical hierarchy. It is worth repeating that the fact that messages reaching the cortex, combine in a single message information about events in the body and the world as well as information about upcoming actions, is not a sufficient explanation, because there are other parts of the brain, like the spinal cord or the cerebellum, that receive efference copies and that must be carrying messages with a similar dual meaning. However, these do not produce perceptual processes to any significant demonstrable extent.

---

[116] In an interesting way, the evidence about the binocularly competing images is more readily related to what the monkeys are perceiving than is the evidence about the forward receptive fields. It is not easy to be conscious of (or to think about what it means to anticipate) the perceptual outcome of our actions, although the loss of such a perception of one's anticipated action can, as we have seen, be perceived by a schizophrenic patient and this forces us to recognize them as cognitive events.

[117] Modulators do not bring such information. The idea of a 'seat of consciousness' in brainstem centres or in the thalamic reticular nucleus depends on the observations of regions from which conscious processes can be turned off or on, generally by modulators, but cannot tell us how or where conscious processes *about* the body or the world are generated. These need inputs from drivers.

This last point may need careful consideration if we look at the very small number of cells in the first visual area (2%) whose responses matched the forward receptive fields recorded by Melcher and Colby.[118] Would this very small number of responding cells be sufficient for a cognitive event? Is there a threshold? This very low level for a primary sensory area, compared to the higher numbers at higher levels of the hierarchy, may be relevant for understanding why even the simplest mammalian brains studied by Kaas (2011) need at least one higher level in their cortical hierarchy.

We have to recognize that the passage of transthalamic messages from one cortical area to another involves more than cortex, and also involves more than just the thalamic gate. Each corticothalamic layer 5 input is highly likely to be accompanied by a motor message carried in one or more branches of the relevant corticothalamic driver input. That is, each transthalamic message at each stage of the hierarchy of transthalamic corticocortical connections will be sending a new message to one or more of the phylogenetically older centres and is capable of contributing to and adjusting the precise nature of the next action.

Here it also should be noted that the transthalamic hierarchical connections also depend on a great variety of subcortical, modulatory pathways that can act on the thalamic gate, although only the layer 6 contributions[119] provide the local, topographic information needed for specific local actions.

The evidence about the response patterns for binocularly competing images, or for the presence of forward receptive fields already summarized, clearly indicates the absence of an easily definable border between actions that are perceived and actions that are not perceived. One part of this gradient from absent to faint to salient, may relate to the number of cells that are active and another to the levels of the cortical hierarchy

---

[118] The percentage reported by Logothetis and colleagues for visual areas 1 and 2 (20%) would suggest that, if the V1 to V2 ratios were similar in both, then only about 3.6% such cells were in V1. Possibly the unique feature of the cortex is the hierarchy of transthalamic monitors that can turn a very weak, barely perceptual event involving 2–4% of the local cells, to one that includes almost half of the local cells.

[119] Direct or relayed through the thalamic reticular nucleus.

involved. However, another that is likely to be relevant, is the activity of the thalamic gate. Is it open or closed? Is it open for all of the relevant cells or only for some of them? And that would vary from one moment to the next, and between lower and higher thalamocortical relays. I suggest that it is a mistake to look for one seat for conscious processes anywhere in the brain.

Consciousness is not an entity that is likely to be localizable, it is not an object, but refers to a broad range of personal experiences. We are looking at patterns of neural activities that relate to particular such experiences.[120]

13.3 The parts of thalamocortical organization that may prove crucial for generating conscious processes

13.3.1 General considerations

A neuroscientist can understand how some of the messages that pass to the cortex relate to the structures of the world. That is, we can read the messages even if we cannot interpret them in terms of the private first-person experiences. We may be able to deduce that certain messages represent a red part of the world to the organism under study, but this tells us nothing about the organism's private sense of seeing red; it tells us about the physicist's world. If we know that at a certain point, when a rat is running a maze, it is aware of which way the head is pointing, is seeing the major relevant landmarks, and is perhaps associating some emotional experiences with this particular situation, these are all facts about the world that we, experimentalists, may be able to read in the messages. They are not facts about what the rat is experiencing when it is looking at a red landmark to the right and anticipates a chocolate feast. The private nature of the redness and of the anticipation of the chocolate is missing from the messages that the physiologist reads.[121]

---

[120] See footnote 5 in Chapter 1, for the reference to Michael Billig's book, *Learn to Write Badly*.

[121] Although it may seem easy for the scientist to imagine the rat's experiences at any one stage.

It will not surprise the reader to learn that what follows in this last part of the book will not include an identification of the relevant crucial feature(s) that can generate specific conscious processes. However, we can work towards this. We can look at the features of thalamocortical circuitry that may play a role and whose exploration in the future may well provide an opening for observations that will lead to useful answers about the neural processing, almost certainly in the cortex, involved in generating conscious experiences.[122]

What properties of the thalamocortical circuits are likely to play a role in changing a sensorimotor response with no perceptual outcomes to a perceived experience? I will argue that they are likely to include (1) the capacity of cortical activity to anticipate an organism's own actions; (2) the hierarchy of cortical areas serving as monitors, each of which monitors the motor outputs of lower levels and can also, itself, contribute to the control of the thalamic gate through its layer 6 outputs; (3) the thalamic gate; and 4) the motor instructions that can be issued from any part of the cortical hierarchy and that can serve to contribute to the ongoing activity being monitored.

These features allow for a temporal sequence progressing from one perception to the next, providing the continuity of experiences, with anticipated actions matching anticipated reafferences which in turn generate actions that are also anticipated. We have learnt to expect this sensorimotor flow from the world.

These properties can also be linked to the increasing percentages reported from lower to higher visual cortical areas for cells responding to binocularly different images or showing responses for forward receptive fields. However, we do not know how the increases are produced in the cortex. These four items listed in the previous paragraph exclude one important and probably crucial factor: the basic connectivity patterns of the cortex itself. In order to understand how conscious processing is generated in the cortex and increased from one level to the

---

[122] In Adrian's terms, cited at the beginning of Part IV, to 'show that the two events [the neural processing and the thought] are really the same thing looked at from a different point of view'.

next, we will need to know about cortical processing in terms that are currently not defined. We will need to learn how the first few cells active during conscious processing are generated in the primary sensory cortical areas,[123] and then to define the processes that generate more such cells at higher levels.

The four features, acting together (almost certainly not one or the other on its own), may prove to be a combination that is important for increasing the numbers of responsive cortical cells from the lowest levels to their fuller role at higher levels.

MacKay (1982, page 291) has argued that:

> From the standpoint of information engineering analysis, the brain of a *cognitive* [italics added] agent requires at least two levels of organization, one of them concerned with maintaining the vast repertoire of conditional readinesses to reckon with the field of action, and the other ('supervisory') level to be responsible for the continual on-going adjustment of priorities and criteria of evaluation....In an important sense, such an information system becomes its own programmer.

This suggestion that the brain of a cognitive agent required a hierarchy with one level supervising another was not MacKay's only account of some such hierarchy (see MacKay 1956), nor did it stand alone. Miller et al. (1960) had written about a plan for guiding action as a hierarchical process able to control action sequences.

MacKay's comments about an information system becoming its own programmer also have an antecedent,[124] which, translated from the original German, reads: 'Konrad Zuse, who, in the 1940s, in all probability built the first modern computer in the sense already discussed in section 1.2, said the following at a lecture at the ETH in Zürich in 1992: 'The idea of programs that can act on themselves, was for me at the beginning rather spooky [weird, uncanny]. Until then, one could get a good overview of the instruments Z1–Z4 ... As soon as I allowed

---

[123] The best place to start may well prove to be experimental recordings of relevant neural pathways while a new perceptual skill, such as viewing the world through inverting lenses, or acquiring a new sense of directional hearing, is being learnt.

[124] Böszörményi (1997, page 115), I thank Fritz Sommer for telling me about this publication.

for the influence of the computed data back upon the program itself – this needed only a small connection [wire?], which linked back from the calculations to the program [programwerk ... the program's record?], this control [this overview of Z1–Z4] is no longer available. I had a great respect for this small connection [wire], because I realized [I had a foreboding?] that Mephisto was standing behind me ... With that one can actually do the strangest things.'

It looks as though at this point we have to deal with MacKay's suggestion, already in the title of his article, that when one level of a program can act on another, lower level, a 'cognitive agent' can be generated without any extra that might imply dualism or a separate soul. This creation of a cognitive agent without dualism would also, of course, have attracted Mephisto.

In order to explore this further, in the future we will need to look at some specific neural mechanisms that relate to specific cognitive processes. We will need to look at the limited relevant details we know about the cortex and use these to define our areas of ignorance. A good example may be the binocular conflict studied by Logothetis and colleagues and discussed earlier in this chapter. The published experiments show us what the sensory pathways are doing, but we know nothing about the 'relevant supervisory mechanisms responsible for the ongoing adjustment of priorities and criteria of evaluation', which presumably are controlling the binocular switch when it occurs. These are likely to be outputs coming from layer 5 of one or more visual cortical areas, which, at a crucial point, change the balance of controls of the images seen by the eyes: perhaps the focus, the pupillary size, or the microsaccades. This will undoubtedly prove challenging and it may be necessary to devise some simpler experimental models, but in the long run, experiments such as this may well provide a road to empirical knowledge about the role our brains play in producing our cognitive capacities.

In the rest of this section I will first enquire about the extent to which we can treat the brain as, at least in part, functioning, in MacKay's terms, as its own programmer, and then consider the requirements outlined by him.

We can treat the brain as including two parts: the phylogenetically old parts of the brain which have proved capable of organizing the complex lives of many of our vertebrate ancestors, and the phylogenetically newer parts of the mammalian six-layered cortex, upon which our conscious lives depend. Possibly it is only the newest, mammalian feature, the cortical hierarchy, rather than the cortex itself, that will turn out to be important here.

The many non-thalamic branches of the layer 5 corticothalamic axons, some of which are illustrated as the motor components alone at the beginning of Chapter 4, and some of which are shown in Chapter 9 in Figure 9.1, represent the vast repertoire of conditional readinesses, required by MacKay's proposal. They innervate many subcortical motor centres, which are concerned with contributing to these readinesses of the phylogenetically old brain for action. That is, the several layer 5 motor branches innervating the phylogenetically old motor centres can be seen as together providing the conditions needed for an action to be successful in producing the anticipated forthcoming perception. They represent the new parts of the brain linking back to the older parts.

The supervisory level is represented by the hierarchy of cortical centres and includes more than the one level needed in MacKay's proposal. There are several levels, each monitoring the motor outputs of lower levels, each receiving (or not receiving) this information through the thalamic gate and each also able to contribute to the control of the gate and to the actions of the relevant motor centres.

MacKay (1982, page 291) also states:

> [T]he flow pattern of supervisory activity offers a natural correlate for the flux of the subject's conscious experience.

This has each sensorimotor event generating the next and suggests that the temporal sequences of sensorimotor inputs that are passed up the cortical hierarchy when a novel situation is encountered are relevant for blocking or generating a conscious experience.

Consider an infant's first visual experiences, or a young shaver's first use of a mirror; or consider, in an adult, the use of prisms that displace the visual image, or attempts at recovery after sensory losses or misconnections: these will at first generate sensorimotor sequences in which the reafference fails to match expectations. That is, the 'flux of the subject's conscious experience' is lost. At first the new sequences make no sense. There can be no successful perception. The descending motor branches must contribute to a change in the sensorimotor temporal relationships at lower levels until the expected flow of matching temporal relationships is once more established. One can think of the higher cortical levels interrogating inputs from lower levels and adjusting motor outputs accordingly until the flow of the expected relationships is once more established. It is only when the flow pattern of the inputs and outputs meets expectations that we can expect that a comprehensible conscious experience can be generated. One would anticipate that this is generated first at the lowest levels and then passed up the hierarchy.

At each level, the nervous system needs to interrogate the sensorimotor relationships passed up by lower levels and ensure that the flow of the perception, the anticipated action, and the matching reafference can be re-instated as an uninterrupted sequence, each one forming a part of the flow of our successful interactions with the world. This can then be expected to provide the conditions needed for the cortex to generate a conscious experience.

For any one cortical area, if one wants to look for the neural events that accompany the generation of a conscious experience, it may be worth studying the activity in the thalamic inputs to the cortex and the outputs from layer 5 as new perceptual skills are learnt. The earliest experiences of a newly displaced sensorimotor relationship can be expected to lack the smooth flow of established sensorimotor skills, MacKay's 'flux', and therefore to lack any significant, comprehensible cognitive experience. One should expect that flux to be reinstated as the new skill is acquired, and in practice, the relevant neural events should be demonstrable, providing a psychophysical parallel between neural and perceptual functions. It should be clear that at present, the actual cortical processes that can produce a cognitive event, or increase its impact as it is passed up the hierarchy, remain as undefined and mysterious as the origins of consciousness appeared to Adrian in the preface to his book.

13.3.2 **The role of the thalamic gate**

The thalamic gate is important because it can control what MacKay calls conditional readinesses, the contributions that the layer 6 outputs and other modulators can make to the transthalamic corticocortical communications *about the forthcoming motor instructions* on the one hand, and that the layer 5 cortical outputs can make to subcortical motor centres on the other. We need to learn more about messages that pass from lower cortical levels to higher levels: how do the direct corticocortical pathways, which would have little to do with the conditional readinesses, compare to the transthalamic pathways, which carry information about the instructions that establish these readinesses? How do the actions of the thalamic gate relate to the ways in which messages pass from a lower to a higher cortical area, presumably having a potential to increase the level of consciousness with each passage?

For example, if this is a new visual relationship that is being learnt, then it would be important to know about the activity of layer 5 and 6 cells in several of the higher visual areas. If this activity could be recorded together with the functional effects produced by the layer 5 outputs to identified lower motor centres then one would be identifying actions that relate to *control* of the relevant conditional readinesses. These actions for the layer 5 visual pathways could be mainly gaze shifts or eye movements, changes in the accommodation of the lens or the size of the pupil. They would be actions that could re-establish the flow of the cognitive process. The basic aim would be to record the patterns of cortical events and their actions on the relevant body parts and to distinguish a perceived event from one that was seen but was not perceived, tracking each message up the cortical hierarchy. We need to understand how sensory messages can affect cortical actions that, in turn, can be related to the specifics of the perceptions.

Comparably, recordings from layer 6 cells would provide indications about one of the ways in which the control of the thalamic gate is changing as a new sensory relationship is being learnt. In the long run, other contributions from the many pathways that control of the gate would, of course, need to be studied as well, including the many modulatory components that relate to mood, attention, and so on

This is clearly too vague and extended to be a single research proposal; it aims to focus on the nature of the thalamocortical inputs (burst or tonic), on the outputs from cortical layers 5 and 6, and of the many modulators to start a search that may lead to differences between perceived events and events that are not perceived. No matter how successful such proposed investigations may prove to be, they will not reveal a seat of consciousness, nor can they show us what a conscious process, a perception, actually is in terms of the subject's own experiences: for the neuroscientist it is the successful creation of the flow of the sensorimotor sequences.

13.3.3 Neural activity that anticipates an organism's actions

In order to identify anticipatory activity for any one driver input to the thalamus it will be important to know the actions produced by the non-thalamic branches of that input (see Chapter 10, Figure 10.1). For example, for any one retinogeniculate axon we know almost nothing about the action to which it is contributing with its tectal or pretectal branches.[125] Further, there are corticothalamic layer 5 driver axons that go from several visual cortical areas to the pulvinar (Rockland 1996, 1998; Guillery et al. 2001). Many of these also send branches to the pretectum and tectum (Harting et al. 1992). However, the specific actions of any one of these midbrain branches are undefined, although they must be contributing to active vision, probably each in a distinctive and definable way. Similarly, we have limited information about the actions to which the spinal branches of specific segmental afferents (tactile, joint receptor, nociceptive,[126] etc.) destined for thalamus contribute. For most thalamic driver inputs, we know nothing about the motor action of their non-thalamic branches. Similar questions can be asked about the actions

---

[125] In part this is because these, especially the tectal branches, have often been treated as sensory inputs. But it is also because almost none of the motor actions that can be assigned to tectal inputs, in general, can be linked to the thalamic branches we are considering here.

[126] Pain receptors.

of their motor branches for almost any other layer 5 outputs to the thalamus. If there is evidence for an anticipation of a reafference, that is, an anticipation of a new sensory input, corresponding to the new message produced by a movement (comparable to a forward receptive field) this could perhaps be considered as a way of distinguishing a cortical event that is perceived from one that is not perceived, provided there were enough neurons demonstrably active at higher levels of the cortical hierarchy.

Once an animal has learnt to anticipate a new sensory event, the fulfilment of that anticipation (the recognition of the expected reafference) plays a role in turning a sensory event into a perceived event: 'the flux of the subject's conscious experience' referred to by MacKay.

Quite apart from the focus of the present chapter, observations about the functional properties of the corticofugal pathways would, in themselves, provide information useful for understanding more about the relevant cortical areas than we generally know at present. It is probable that for a first-order thalamocortical pathway the motor actions arising from layer 5 of the relevant cortical area are likely to produce relatively minor adjustments; for a well-practised action,[127] the movements will be accurate and corrections will rarely be needed. However, for new actions that are being learnt, as, for example, early responses to reversing lenses, or an infant seeing an object for the first time, the sizes of the needed corrections are likely to be greater and easier to identify by a physiologist. They are likely to involve outputs to the pons for cerebellar adjustments as well as outputs to the striatum and superior colliculus (see Chapter 4). One conclusion that follows from these suggestions is that perceptions depend on well-learnt sensorimotor relationships, and the reports of learning to live with reversing lenses would appear to provide some preliminary support for this idea.

---

[127] The extent to which well-practised actions differ in this way from regular actions that are not extensively rehearsed, may prove to be important for interpreting many behavioural experiments.

It has been mentioned that visual forward receptive fields are more commonly found in higher visual cortical areas than in first- or second-order visual sensory areas. The extent to which this depends on the repeated monitoring of motor outputs, one level after the other, collecting information about expectations and thus strengthening the sense of a perceptual presence remains to be defined. Probably the anticipatory functions would play a minor role for a well-rehearsed action, which would be primarily controlled by lower centres. The perceptual role might well be more important, and would often depend on an unanticipated event and a thalamic wake-up call that might increase the perceptual presence of the event. Overall the aims of future studies would be to relate the anticipatory neural activity to the perceptions and to define how perceptions depend on actions.

One way at looking at the implications of this section is to think of conscious processes as being a temporal sequence of neural events based on anticipated, and therefore recognizable events.

13.3.4 The role of the hierarchy of cortical monitors

The fact that there is a hierarchy of cortical areas each monitoring the motor outputs of lower cortical and of non-cortical cell groups may prove to be one of the most strikingly original developments of the mammalian brain. Each cortical area has its own outputs to lower motor centres, commonly to several of them. Some of these go to centres like the superior colliculus or the red nucleus that have direct or ready access to spinal ventral horn cells or cranial motor nuclei (see Chapter 4, Figure 4.1), whereas others go to the striatum or the zona incerta. These structures, in turn, have inhibitory or modulatory actions, not only on the lower motor centres but also on the thalamic relay cells that feed the cortex. The striatal pathways that go through the globus pallidus and substantia nigra to brainstem motor centres and the zona incerta can, similarly, also act on the thalamus and also on the brainstem centres. They can serve for continual ongoing adjustment of priorities in MacKay's sense.

13.4 **The nature of the anticipated motor instructions**

In order to understand how actions relate to perceptions we need to know the actions produced by the layer 5 motor branches, how the strength or the dominance of the actions relates to the cortical hierarchy at any one moment, and how these motor discharges can contribute to conditions where reafferences match expectations.

Much depends on the motor instructions that any one cortical area is sending to the phylogenetically older subcortical structures. I mentioned in Chapter 4 the extent to which the rich cortical inputs to the superior colliculus can dominate the activity of the collicular cells and also that, through the corticostriatal pathway and the substantia nigra, the cortex can modify the activity of the colliculus. The cortical inputs to other lower motor centres of the brainstem like the inputs to the cerebellum or the cortical input to the red nucleus have to be considered in a comparable light as being able to dominate, silence, or free up the activities of one or another of the lower motor centres.

If we now ask what MacKay had in mind when he wrote about the 'important sense [in which] such an information system becomes its own programmer' we have to distinguish which parts of the brain represent the programmer and which the program. The discussion so far suggests that, although the distinction cannot be an entirely rigid one, it makes rough sense to treat the phylogenetically old brainstem and spinal pathways together with the basal ganglia and the cerebellum as the established program, and to see the cortical pathways to these centres as providing new (learnt) data that have been processed and can be sent back to the old parts of the brain. However, in view of the very low number of responsive nerve cells reported in the experiments on binocular competition or the forward receptive fields for the primary visual area (see earlier in this chapter), and the related observation that a two-level hierarchy may the present in the earliest mammals, it is possible that the crucial newest development feeding back to the phylogenetically old brain is represented by the second level of the cortical hierarchy. In that sense, MacKay's proposal relates very closely to what we currently know about the brain: providing a sketch of the basic connections on the basis of which we can use our brains to interrogate the world, showing where our perceptions match our anticipated actions and reafferences, and indicating how we can expect to find the best relationships for generating the continuous sensorimotor flux on which our conscious lives depend. One could read this as the neural activity corresponding to the unconscious conclusions that Helmholtz saw as the basis of our cognitive interactions with what appears to us as a real world.

At present we know all too little about the relevant corticofugal pathways from most higher cortical areas, and where we do know the connections, as, for example, for the corticopontine or corticostriatal connections, the known details of the actions that these pathways produce or modify at lower levels are not sufficient for us yet to understand how any one cortical area, other than the motor cortical areas themselves, may be contributing to the subcortical centres in ways to alter sensorimotor relationships.

There is an intriguing question that relates closely to the earlier question as to what it is that the mammalian six-layered cortex provides for a mammal that is not available for our non-mammalian ancestors. The view that the cortical inputs to the phylogenetically old motor centres of the brain are a new addition that serves to program (or re-program?) the older motor centres, raises an interesting question. What created the mammalian need for this new input to old mechanisms?

One possible answer concerns the relationships established by mammalian offspring. They need to establish interactive relationships with their litter mates in a very close environment that relates to their mother and their nest, cave, or other home environment. This unusually close interactive, social environment, not present in premammalian forms, will have made unusual sensorimotor demands on the earliest mammals as soon as they started being parts of relatively large litters, and an early addition of even a small hierarchy of corticothalamocortical pathways with branches for modifying motor actions, necessary for responding to the social demands of the family, may have proved a useful addition for successful interactions with litter mates, providing a new set connections for 'evaluating conditional readinesses to reckon with the field of action'.

13.5 **The role of the layer 5 cortical cells**

Earlier I indicated that a key feature, currently unanalysed, is the actual nature of the cortical processing that occurs at each level of the hierarchy, establishing or strengthening cognitive events. The layer 5 thalamocortical cells with their branched axons and complex dendrites must clearly be playing some role. The evidence available about the complexity of the several inputs to the different parts of these cells and the nature of their outputs may well prove relevant for understanding events relevant for the generation of conscious processes.

The cortex needs a cell that can respond to and adjust the intervals between the arrival of a message about a sensory event with its corresponding efference copy, and the subsequent arrival of the reafference (and *its* efference copy). The precise timing of this sequence of afferent events is basic to the cortical reception of the continuous flow of sensory experiences that characterizes our conscious lives. It needs cells that can estimate intervals between events and adjust them.

In the somatosensory cortex of rodents, where cells respond to whisker stimulation, the layer 5 cortical cells provide a higher-order input to the thalamus[128] (Killackey and Sherman 2003; Theyel et al. 2010; Mease et al. 2016) for relay to higher cortical areas that can monitor the flow of the incoming sensorimotor activity. The axons of these cortical cells also send motor branches (Deschênes et al. 1994; Veinante et al. 2000) to the midbrain. Their motor actions have not been defined in detail but the anatomical sites of their termination suggest a role in the control of eye movements, pupillary and lens control and head movements. This looks like a necessary link between the whisker movements and vision. How these relate to the control of the relevant whisker movements themselves is not known, and the possibility that these corticofugal axons also play a role in the control of whisker movements may prove relevant.

The outputs of these layer 5 cells have been recorded as short bursts of action potentials (Larkum et al. 2001), the sort of message that is

---

[128] The higher-order nucleus that in rodents is called POm.

likely to be an error signal or a motor instruction rather than a sensory message for further cortical processing.[129] The relevant layer 5 cells have three distinct input zones (Larkum et al. 2001; Sakmann 2017), which demonstrate plastic properties and interact to produce the outgoing action potentials. The timing of these several inputs in relation to each other will be crucial in adjusting the sequence of sensorimotor events occurring between the tactile and the visual explorations that a rodent is carrying out. Early in life this timing must be learned, and later in life it must be accurately maintained and adjusted to sustain the flow of conscious processing. Whereas these **pyramidal cells** have been described as coincidence detectors they may also serve to adjust the timing of the inputs in relation to the outputs, serving as delay estimators.

That is, the layer 5 bursts can be expected to provide corrections of the precise output timing when this is not producing a smooth flow of sensorimotor interactions. At the same time these bursts are also informing higher cortical areas that something needs to be corrected, is being corrected, or needs no correction. That is, these layer 5 cells and their inputs may provide some important clues for a deeper understanding of the cortical role in establishing and maintaining the flow of perceptions and actions.[130]

I have presented several areas of ignorance that may perhaps prove relevant for understanding how the cortex can generate an awareness

---

[129] Since these whiskers normally act as a part of a group, sweeping in serial order across the nearby environment, whiskers that lack the responses to such a normally expected sequence may well be sending out error signals.

[130] Ocaña et al. (2015) describe the pyramidal cells in the relatively simple cerebral cortex of the lamprey sending motor outputs to subcortical motor centres. They appear to play a role in the control of motor centres not unlike that played by mammalian layer 5 pyramidal cells. The crucial difference is likely to be the role of the thalamic branches for relay of information to higher cortical areas that characterizes the mammalian layer 5 cells. It looks as though the thalamocortical branch has been added to the phylogenetically older motor output. We need to think of the mammalian replication of cortical areas and their hierarchical interconnections as one item that generates the power of the mammalian cortex, and then to ask how the role of the more complex six-layered mammalian cortex relates to this.

of events and actions. However, there is another way to look at these areas of ignorance and that is to see them as significant if we want to understand what any one cortical area can reasonably be expected to be doing. That is, research into these questions should be seen not only as a way to learn more about the cortex in relation to conscious processing, but also as contributing fundamental information about the functional capacities of any one cortical area.

Chapter 14

The neural origins of a sense of self with a brief note on free will

14.1 Summary

This chapter will be limited to a brief discussion of a sense of self, based on points that relate to the neural mechanisms discussed in previous chapters, in order to illustrate some important differences between the standard view and the integrative sensorimotor view of the self. I will treat the sense of self as produced by a temporal sequence of related perceptions that lead to 'unconscious conclusions' about the self. I shall address two major questions: (1) does our ability to anticipate our own actions contribute to a sense of self? Providing a clear 'yes'. (2) Is our sense of self limited to the boundaries of our own bodies, and to the neural signals we receive from them? Providing a clear 'no'. I will briefly address our sense of the self as agent, recognizing that this extends significantly beyond the content of this book.

14.2 The self as based on perceived actions

When Descartes argued 'I think therefore I am' he established a reason for taking his sense of self seriously. There had to be someone doing the thinking and it is unreasonable to consider that it might be someone else. Once we recognize that we commonly anticipate our own actions, and that there are pathological situations in which people assign their actions to others because they can no longer identify their actions as their own, the same argument can be extended beyond our thoughts to our actions. The abnormal condition, reported particularly for patients suffering from schizophrenia, of actions and thoughts assigned by patients to other people or agents, illustrates the extent to which the information available to us about our forthcoming actions contributes to our sense of self *on the basis of the perceptions we have about our own actions.*

It is important to recognize that this use of 'self' treats the self as a first-person agent: I act, therefore I am. It is not an extra entity required by a dualistic view, comparable to a mind or a soul. It is, like our knowledge of the world, a construct based on our perceptions, in the sense used by Helmholtz, it is another unconscious conclusion.

We saw earlier, in Chapter 10, that the responses of nerve cells can anticipate an action of the self in a way that actions of others cannot be anticipated. No matter how well we know someone else, or how good our reasons for believing we know what they will do next, our perceptions of our own actions differ in kind from anything we can anticipate or experience about another's actions. In neural terms, the efference copies that carry information about our forthcoming actions show that our perceptions of our own actions are produced by a pattern of neural activity that clearly distinguishes these perceptions from our perceptions of the of the world, including the actions of others. In view of the role that thalamocortical hierarchies play in consciousness, discussed in Chapter 13, I will treat the relevant neural activity as cortical and as likely to involve higher cortical areas. Conscious anticipations of our own actions serve as first-person experiences, not available to others, and contribute to our sense of self. Perhaps, most tellingly, we can notice when we lose the ability to identify our actions as our own.

It is the anticipation followed by the execution of many (certainly not all!) of our anticipated actions that plays a significant role in our sense of self. Damasio (2006, pages 239–40) wrote: 'The collective representation of the body constitute [sic] the basis for a "concept" of self, much as a collection of representations of shape, size, color, texture and taste can constitute the basis for the concept of orange'. Dehaene (2014, page 24) wrote about the perceived nature of the self: 'In my view, self-consciousness is much like consciousness of color or sound'. These views are based on the standard sensory-to-motor view that includes pure perceptions.

That is, these authors treat self-consciousness as merely another pure perceptual process about an object, the same as any other in the world. When, on the basis of the present account, we ask what is perceived as regards the self, we have to include the anticipations of our own actions. These form an essential part of our *sense of* self. Here, it may be useful to stress the verb that has us sensing our actions and by anticipating them, recognizing them as our own, and to diminish the noun that seems to imply something resembling a fruit or a vegetable. The integrative view of our interactions with the world differs from the standard view and on the former, our views of ourselves differ from any views we may experience of colours, fruits, or vegetables on a plate. Although the anticipation of our own actions may be a crucial difference, another neurally related difference may concern the level of the cortical hierarchy that is involved in routine perceptions of fruits or vegetables relative to those relevant to perceptions of the self. We should anticipate that the highest levels of the hierarchy would often be involved in perceptions of the self, whereas most of our perceptions of fruits and colours might be dealt with at relatively low levels, unless an artist or scientist is studying them. More significantly, in our perceptions of the self, the actions and reafferences provide a continuous timeline between actions and perceptions for our unconscious conclusions about the self, which is only rarely lost by periods of sleep or loss of consciousness.

The self has a history in our experience that fruits, vegetables, and colours lack. This makes our perceptions of the self significantly different from our perceptions of fruits, vegetables, or colours and also different from our perceptions of other people's actions. Yes, we recognize the sun and the moon and the house opposite, our parents, and our children as having a temporal continuity, they have their own histories, but it is not a continuity that we can ever experience in the sense that we know ourselves under normal conditions, where anticipations of actions are followed by actions in one essentially continuous temporal sequence. That is (excluding sleep and conditions of mental pathology) we spend our conscious lives in the continuous presence of the self. The presence may be strong or weak, but it does not go away. In terms of the neural activity we are considering there is something special about our perceptions of the self. Not only are these perceptions based on our own actions; more importantly, they are based on actions that fit our anticipations. That, as I have argued in Chapters 10 and 11, is how they can be recognized as our own actions. They differ in terms of the sequence of neural events, and also differ introspectively, from colours or sounds, apples or radishes, the sun or the moon. They have a unique internal continuity. They play a crucial role in our sense of time.

Kenny (1989, page 9) has written that 'self-consciousness is not possible without language: without language there is no difference between being in pain and having the thought "I'm in pain"'. Here and elsewhere in his book he is treating self-consciousness as a pure sensory experience that needs a description, whereas I have treated it as an experience that involves an action.[131] Can we not know our own actions without language? Can it be correct to consider that a dog might not be able to identify a pain as its own, even when the movements made by the dog significantly alter the pain with each movement? It seems doubtful. The point may be that it does not make sense to ascribe one's pain to someone else, and thus the issue of whose pain it is generally does not arise.[132] However, when the perception is clearly linked to a forthcoming action, an action that can also be (mistakenly) assigned to an external agent as demonstrated by schizophrenic experiences, then the distinction between perceiving an action of the self and an action of some other agent should be evident without the need for language.

It should be possible to establish experimentally whether a rat or a dog can distinguish its own acts from acts apparently not initiated by itself, either because they have been initiated by a physiologist's stimulus, or because the transthalamic pathway, which allows the recognition of an act as self-initiated, has been blocked.[133] Although recognizing an action as initiated by self or other is clearly based on perceptions,

---

[131] The action may well be a thought.

[132] Dickens in *Hard Times* wrote, 'I think there's a pain somewhere in the room' said Mrs Gradgrind, 'but I couldn't positively say that I have got it'. One has to wonder whether this was based on a personal experience of a visit to a very sick person.

[133] It is important to recognize that a comparable experiment about pain does not make sense.

comparable to other perceptions, first person and subject to unconscious conclusions, the unconscious conclusions about our own actions have a force that is as certain and final as most of our other unconscious conclusions about the world, and not in need of linguistic explanations or justifications.

Strawson's (1997, page 406) statement, 'the sense that people have of themselves as being, specifically, a mental presence; a mental someone; a single mental thing that is a conscious subject of experience, that has a certain character or personality'[134] does not translate readily into neural terms, but I have a strong sense that in many ways it is not far from a view of the self I have here tried to sketch in neural terms. Translating the mental to the neural presents many problems. As indicated earlier, we cannot expect to *explain* the mental reductively in terms of the neural, but the effort to attempt descriptive parallels looks worth exploring. One has to recognize that Strawson's statement includes more attributes of the sense of self than I have considered by limiting the discussion to efference copies. The gaps may well be filled systematically by considering other neural events relating to experiences, to character, and to personality, possibly even including some of what is implied by a mental presence. However, the project then becomes open-ended. It is a project more suitable for a saga, a novel, or a long autobiography than a neuroscientific study.

Popper wrote, 'The self in a sense plays on the brain, as a pianist plays on a piano...' (Popper and Eccles 1977, page 494).[135] This is a tellingly apt metaphor, particularly if we think of a pianist who can at times play extremely well and can learn a piece rapidly. However, there is a question about the use of the 'self' here, which may seem to distinguish the agent playing from the organism itself, something that may or may not be intended on the basis of other parts of the 1977 book. I will present it here as not referring to an independent agent but as another way of

---

[134] Strawson is concerned to argue that the sense of self represents a sense of a real mental entity, but from the current point of view, of the self as *represented by* neural activity, we need not follow him along that path.

[135] I am grateful to Andrew Parker for pointing this out to me.

saying that we use our brains as a tool, or of seeing higher levels of the cortical hierarchy (which play an active role in generating a sense of self) 'playing on' the phylogenetically old brain.

14.3 **Is the self represented by our bodies?**

One way of thinking of one's self is to identify the self in terms of one's own body. The body appears to have easily defined borders and is generally clearly under our control in a way that other parts of the world do not appear to be. We can broadly identify the neural pathways that allow us to perceive our bodies and their actions. When I look in the mirror or view photographs I have another way of identifying myself, although this can raise problems:[136] there are serious questions as to who or what this 'self' in the mirror or in the photographs is. My mirror image is not the same as the 'true image' seen by other people; nor do photographs (or movies) generally do a good job of convincing us that this is who we 'really' are, any more than do recordings of our own voices. We can get used to the mirror image, the photographic record, or the recorded voice, but it is not an identification that comes naturally. It is a learnt recognition and is never very convincing.

There are other problems: does our sense of self depend on our clothes? Clearly it does, that is why people dress up for special events. To explore this further, defining ourselves in terms of our naked body is probably not a definition that springs readily to mind for most of us. The perceptions that produce our sense of self on the basis of our body image are elusive. They have to be learnt and they can be inconsistent; they provide a rough guide. One clear point about these perceptions of the self is that they are based on messages that pass along the sensory pathways to the thalamus and are relayed to cortex. They are potentially comprehensible in neural terms. They include not only views of ourselves but also perceptions of our anticipated actions as well as memories of our past actions.

---

136 When I was 11 or 12 during the Second World War, my mother told me that I should never part my hair on the right because then I would look like Hitler. I tried to persuade her that I did not look at all like Hitler in the mirror, but she never got the point.

In addition to a rather variable nature of our sense of self, based on what we know about the body and its actions, we also have to consider suggestions that the self may extend well beyond the body. Clark (2008), in his book *Supersizing the Mind*, argues convincingly that we should expand our understanding of the mind to include many of the tools that we use to aid us. He includes his mobile phone and also a diary that an Alzheimer's patient might use to help find and keep his appointments and so on. He suggests these as examples of objects that extend the functions of our minds. On that basis, Clark writes about an extended, or supersized, mind. In writing here about the self (as opposed to the mind), some of the same issues of defining the borders occur. I started this discussion with our clothes but our tools provide another example and one that is easier to think of in neural terms.

When I use a pencil, the details of the writing are not produced by the movements of my hand but by the movements of the pencil's tip. The pencil's tip plays the role that my finger plays when I write on a misted-up window. The relevant neural connections adapt rapidly to the size and the weight of the pencil to produce the required result, and after very limited use of a tiny or a very large pencil the visual control of the pencil's movements plays a minor or no role. The changed sensorimotor relationships are readily made a part of new movement patterns. I can change the weight and the length of the pencil, the nature and angle of the writing surface, and still write without much learning of the new conditions. The neural mechanisms incorporate new relationships into the necessary control of the movements needed with surprisingly little learning involved.

A perhaps more striking example is the stick used by a blind person. This serves as a sensor, partially replacing vision. A simple wooden stick can serve as a sensor because the neural pathways concerned rapidly adapt to the physical properties of the stick, primarily its length and the distribution of its weight, and in this way the stick can serve to generate perceptions that are otherwise unavailable. The tip of the blind person's stick, a long way from the body, becomes a tactile tool. It serves in a way closely comparable to the tips of our fingers looking for an object on a table in the dark. The stick, the hand that holds it, and the arm together represent a sensory–motor combination that includes the joint receptors and the muscle receptors of the arm and hand, as well as the tactile receptors of the hand and fingers to send messages to the brain, not just about the movements of the body, the arm, and the fingers, but also about the movements of the stick. They provide information to the nervous system about where the tip of the stick is and how it is encountering the environment. The boundary between the body and the world is lost in this interaction. We can readily arrive at unconscious conclusions about the movements of the tip of the stick. The messages that pass to the thalamus include information not merely about the anticipated movement of the arm, but they also include information about the anticipated movements of the tip of the stick. The information that reaches the thalamus about such an unanticipated contact is about a single event that involves the body, the stick, and the world.

The spectacles discussed in Chapter 13, which invert the image of the world or displace it, challenge the brain to generate new sensorimotor relationships between movements of the hands, the eyes, and the perceptions of the world seen through the spectacles. Once the change has been established, the new relationships, dependent on the spectacles, become a part of the self comparable to the relationships that existed before the spectacles were first worn. The fact that the subject eventually becomes unaware of the inversion or displacement, and that a short period of relearning is required after the spectacles are removed, argue in favour of a view that sees the spectacles as forming an integral part of the self in way not strikingly different from the relationships originally established by the biological lens of the eye.[137]

Comparable arguments can be applied to the prostheses discussed in Chapter 12. Clark (2008) has pointed out that when a vestibular prosthesis reports the movements of the head, it can play a role similar to that of the sensory apparatus of the inner ear, leaving the interaction of the vestibulo-ocular and the conscious experience reflex unchanged.

---

[137] The fact that the biological lens inverts the image of the world on the retina has sometimes puzzled people, but once one recognizes that we learn our interactions with the world it becomes clear that the orientation of the retinal image is not important.

This seems like a clear example of extending the self. Once the links between the prosthesis and the movements have been learnt, and the subject is no longer aware of the actual tactile experiences on the tongue as a part of the new experiences, the prosthesis can be regarded as a crucial part of the self.

The idea that the border between the self and the world is arbitrary is hard to deny. The mobile phone or the diary will strike many as close to an extreme,[138] but they serve to illustrate the problem of defining the self as an entity. I long ago watched a very large earth-scoop as it dug up the road outside my home to replace the water mains. It was operated by a man as though it was merely another of his limbs. As I watched his completely confident manipulation of a very large piece of machinery, his eyes on the scoop not on his hands or the levers, I felt that there should be a representation of that equipment not only in the man's sensory cortex but also in his motor cortex. If the self can be defined by its biography, then perhaps such a large earth scoop should be treated as a part of this man's self, provided that he has been operating it for a long period and no longer thinks in terms of the movements of the levers he manipulates but in terms of the movements of the shovel that is digging up the road. Our almost completely unconscious use of the steering wheel when we are driving raises a comparable problem about the border between the self and a car we know well.[139] From the point of view of relating neural activities to our view of the self we can take a rather adaptable view, understanding that well-learnt actions modify the nervous system and through the life of an individual these learnt actions can significantly alter the nature of our sense of self.[140]

If we now recall that the actions discussed in the previous paragraphs all involve thalamocortical pathways that carry anticipations of forthcoming actions, then it will be clear that once we have learnt the use of

---

[138] They are more difficult to accept because rather than simply changing the pattern of the required movements, they require far more complex adaptations.

[139] See also O'Regan and Noë (2001) on the sensorimotor contingencies involved in driving a car.

[140] What happens to the self of a jockey riding in a race? Can it be entirely separate from the horse?

a tool or a prosthesis so that it no longer stands as a barrier between our actions and perceptions, but instead forms an anticipated part of the flow of the sensorimotor relationships, then the tool and its properties become an integral part of our perceived self and play their part not only in our actions but also in our anticipations of actions and perceptions.

Part of the importance and fascination of the dual messages to the thalamus, which inform us of events in the body and the world and which are also telling us about our upcoming responses to those events, is that these messages contribute to our understanding of the difference between ourselves and others. It is me acting, because no one else can be expected to know my actions before I perform them. I act (it is not someone else ... I know that) therefore I am. The strong conclusion that I exist can derive from actions, not only from thinking, as might be concluded from Descartes' statement. From this point of view there may be nothing special about thinking: it may involve some higher levels of the thalamocortical hierarchy, but it can be treated as one of the more complex actions that lead people to have a sense of self, but one that can be more complex and harder to study.

14.4 **The self as agent**

So far, I have been writing about the *sense* of self in terms of perceptions, not the self as an object or an agent. The sense of the self as agent, seen from the point of view of the standard, sensory-to-motor view of the brain has been extensively explored by P Haggard and colleagues (see, for example, Caspar et al. (2016) and earlier publications cited therein) among others. It leads to conclusions quite different (and more puzzling) than those of the sensorimotor view explored in this book, because it does not recognize the anticipation of actions brought to the cortex by the dual messages that the thalamus passes to the cortex, not only from the world and the body, but also from all levels of the cortical hierarchy. The view that we know when we initiate actions because we receive copies of the relevant motor instructions ahead of the actions, is essential for what I have presented as a neural view.

All we can know depends on our perceptions and on our unconscious conclusions based on those perceptions. If we ask about a 'self' as an independent agent that has (or lacks, depending on one's point of view) free will, then one is not discussing just the perceived self, but one is discussing an individual who does things, who needs to be described by a personal pronoun, as part of the world that we know through our unconscious conclusions. I will treat this conclusion as appearing to be far less firmly grounded than those about apples, houses, and people. Our ways of testing this particular unconscious conclusion are significantly more elusive. The agent discussed in this section will be treated as a person, the whole organism, not as dualistic add-on, a ghost in a machine.

It makes no sense to treat the self as an agent and also to question what our perceptions identify as a key feature of the self: the ability to identify our actions as our own because we can anticipate them.[141] This sense of our own actions seems to tell us that we have free will, and when we lose the ability to anticipate our actions, we can seem to lose our sense of free will, our sense of being in control. When free will is discussed in these terms we can be pretty sure we have it. When it is discussed in terms of the physical laws that govern our actions, then it is sometimes concluded that these laws are inexorable, and that, therefore, we do not have free will (recently summarized by Baggini (2015)).

This is a strange argument and I will consider it only briefly, recognizing the argument as going some way beyond my stated aims of studying the brain as a tool. It would make more sense to argue that if our bodies did not obey the laws of physics we would most likely not have free will. Claude Bernard stressed that the stability of the internal environment (the *milieu intérieur*) is the condition for the free and independent life (Bernard 1974).

We would probably never have evolved to our present form under conditions that are not governed by the laws of physics. Nor would we have stood much of a chance if occasionally the laws were varied in

---

[141] Harris (2015) wrote that Hume reduced the status of the will to 'a kind of epiphenomenal accompaniment to action: Hume described it as, merely, "the internal impression we feel and are conscious of, when we knowingly give rise to any new motion of our body, or new perception of our mind"'. On the basis of this book's contents, this passage does not look like an epiphenomenon but more like the perceptions that anticipate our actions.

accord with the suggestion of one or another philosopher (or scientist acting as philosopher) attempting to prove that somehow our actions are not governed by the laws of physics. We are organisms whose properties depend on our genes and on our environment: on nature and nurture[142] and on the ways in which our brains interact with these starter sets. These contributions to our make-up define who we are. For those who worry that the laws of physics leave us with no free will, something else, some other component outside of those laws is proposed: the real self, the mind, or the Will, are thought to be beyond the contributions made by nature and nurture. Some have invoked chance or quantum uncertainty as being the needed ingredient, but these do not provide an opportunity for us, for some entity acting for or within us and not governed by the laws of physics to change a synapse, grow or shrink a dendrite, increase the number of **dendritic spines**, or move microtubules in order to generate neuronal changes needed for learning or controlling an action. From a neuroscientist's point of view there is no extra component or force to do that. The brain is sufficiently complex to serve as a tool for resolving conflicts, and the way any one conflict is resolved helps to define who we are and who we are likely to become in the future. No, the brain does not resolve the conflict, it serves as tool that we use for resolving the conflict. Without it we can do nothing. With it we have the most wonderful choices, including that of learning more about the nature of this tool.

14.5 **A final conscious conclusion**

Our sense of being in control is as reasonable and practically valuable on the same basis as, in our practical lives, we treat the world as real and it pays off. However, as suggested in the previous section, it may be a bit less firmly based. Our *sense* that there is a real world out there, with trees and meadows, cows and apples, pencils, paintings, and people is incredibly useful. It would be stupid to jettison that

---

142 I am not arguing that nature and nurture are not factors that severely limit our freedom to choose particular actions. The choices we have depend on our place in the world, on the brains we have, and on how we use them to interact with the world.

sense of a reality even when we know that the reality is nothing more than an 'unbewusster Schluss' (unconscious conclusion). In exactly the same way it makes no sense to ignore our sense of ourselves as freely acting agents.

The complexity of the hierarchy of corticothalamocortical connections contributing to and interacting with the several interacting subcortical motor centres provides ample opportunities for decisions made at one level to be over-ridden from a higher level, for mistakes to be made, and for 'errors' to be corrected. Once we learn more about the ways in which the hierarchy of cortical areas and their widespread motor outputs relate to the control of our actions, we may be able to read the activities of the brain in a more telling way. It is not nonsense to think that there are higher parts of our brains that can control lower parts and that the higher parts may well act on the basis of learnt moral and ethical precepts. What we have learnt may depend on our intelligence, our genes, and our environment. Judgements about what we should have learnt, given our circumstances, are judgements that lawyers and moralists need to make, and occasionally the laws are modified on the basis of what is known about the brain's functions. It may well become evident that the interactions that neuroscientists can reveal, interactions between the several parts of the brain, the phylogenetically ancient as well as the newest bits of the frontal and temporal lobes, can serve to show us ourselves more clearly as, indeed, we see ourselves. That is, as captains in control of the ship, of our 'self'. We may start to approach a situation where the account that neuroscientists are able to give of neural processes matches a little more closely what people are really like: complex organisms that do surprising and unexpected things but things that can always be traced to physical causes.

Neuroscientists have a contribution to make to understanding who we are and how we function, but at present this is still relatively limited and there will always be much that is beyond the reach of neuroscientific analysis. Yes, we can understand that damage to certain brain areas produces personality changes and that a patient cannot be held responsible for these. There has been a tremendous change in our attitudes to mental health treatment over the past 200 years.

Some of it is due to a better understanding of the brain, but much of it is due to a better understanding of people. I recently met an eminent contributor to studies of dyslexia who claimed to know very little about the brain. We do not know the neural basis of dyslexia or autism or many other personality disorders but that does not mean that they cannot be understood at a level that can help the patients and their families to a significant extent and can serve to modify laws and treatments. Neuroscientists can provide pointers and clues as to how certain brain regions relate to certain functions. There is a long way to go, and there is an enormous terra incognita, areas of ignorance that it will take a long time to explore adequately. We can expect to learn much about how the brain functions and relate that to who we are and how we act. But if we want to understand people, neuroscience will always be limited, just as our ability to understand people will always be limited. It takes a large slice of our cultural heritage to understand people. We need to look at plays, novels, poems, and detailed studies in many different disciplines ranging from history to psychology and from the social sciences to pharmacology. We cannot expect that neuroscience, no matter how advanced and sophisticated, will show us ourselves either as we see ourselves or as we 'really' are, whatever meaning we decide to give to this.

Glossary

Note that this glossary is strictly limited to the information needed for the text of this book. Much more could be said on most of the items.

Action potentials: very brief changes in the local electric potential across the membrane of an axon, which pass along axons. These action potentials form the basis of the messages passing from a nerve cell to its axon terminals. Recordings of action potentials are shown in Figure 2.3, which illustrate the way in which the dynamics of the action potentials can represent information about events in the body and the world to an investigator.

Afferent: an afferent message is one coming into the nervous system from the body or the world.

Anterior cingulate cortex: an area on the medial surface of the cerebral cortex, labelled 1 in Figure 6.1. It receives messages from one of the anterior thalamic nuclei, labelled 8 in Figure 6.1, and is discussed in Chapter 6. Currently we do not know the nature of the messages that this cortical area receives from the thalamus.

Aphasia: motor aphasia is a loss of the ability to produce speech, even when comprehension of speech is not affected; sensory aphasia is a loss of the comprehension of speech. *See also* Broca's area.

Auditory pathways: neural pathways concerned with hearing.

Axon: this is the single slender process by which one nerve can communicate with another over considerable distances by the passage of action potentials (*see* action potentials). The axon differs in appearance and functions from the several other processes of nerve cells, the dendrites (see Figure 2.4).

Axon hillock: the part of the cell from which the axon arises as a slender structure that gradually thickens. This is the initial segment of the axon, where the action potentials originate.

Basal ganglia: a group of neural structures that lie centrally in the brain (labelled BG in Figure 4.1). They play a role in the control and the initiation of movements. They receive inputs from most areas of the cerebral cortex, and also from several thalamic nuclei. They have important outputs to the motor centres of the brainstem, some of which are shown in Figure 4.1.

BOLD signal: these are signals recorded by measuring changes in local oxygen levels in the blood vessels of the brain. These blood oxygen level-dependent signals reflect the level of neural activity in a local area of the brain.

Brainstem: a part of the brain that lies between the brain and the spinal cord. It includes the midbrain, the pons, and the medulla in that order from the brain to the spinal cord. Many important centres lie in the brainstem and many important pathways pass through it (see labels 12, 13, and 14 in Figure 1.1(b)).

Broca's area: Broca (1863) described an area of the cortex (labelled 4 in Figure 1.1(a)) damage to which produces a motor aphasia (*see* aphasia).

Burst mode of firing: a pattern of closely spaced action potentials produced by a thalamic relay cell when it receives a strong input after a short period of silence. The bursts do not represent the dynamics of the message, only the presence of a message that may need attention. Once the bursts reach the cortex, the cortex sends excitatory inputs back to the thalamus, serving as a 'wake-up call', changing the burst mode of the thalamic cell to the relay mode (see Figure 8.7).

Caudal: towards the tail; that is further from the cortex and nearer to the lowest part of the spinal cord.

Caudate nucleus: one of the major cell groups of the basal ganglia (labelled CN in Figure 4.1) (*see* basal ganglia).

Central nervous system: this includes the brain, the brainstem, cerebellum, spinal cord, and the olfactory bulbs. It may also include the pituitary and the pineal glands. The retina is an embryonic outgrowth of the brain, but is usually not treated as a part of the central nervous system.

Central sulcus: a cerebral sulcus (labelled 1 in Figure 1.1(a)), which separates the precentral from the postcentral gyrus (labelled 2 and 3 in Figure 1.1(a)).

Cerebellum: a large part of the brain concerned with the control movements. It lies below the occipital lobes of the cerebral hemispheres (labelled C in Figure 1.1(a, c)).

Cerebellar peduncle: a large bundle of fibres that links the cerebellum to the brainstem. There are three on each side. The largest, the middle cerebellar peduncle, carries messages from the cerebral cortex to the cerebellar cortex and is labelled as 7 in Figure 1.1(c).

Cerebral cortex: a layer of nerve cells and their processes that covers the surfaces of the cerebral hemispheres (see Figure 1.1(a–e)).

Cerebral peduncle: a large bundle of nerve fibres that carries messages from the cerebral cortex to the brainstem and spinal cord (labelled CP in Figure 4.1).

Cilium: a tiny, microscopic process, that is often mobile, either moving small organisms, or creating a current around a cell. Some are not mobile and these often carry sensory receptors.

Cingulate cortex: an area of cortex on the medial aspect of the cerebral hemispheres. It is divided into three parts: anterior cingulate cortex, posterior cingulate cortex, and retrosplenial cortex (labelled 1, 2, and 3 in Figure 1.1(b)).

Column: used to describe a narrow column of cells that includes the whole thickness of any one cortical area, as in Figure 1.1(e). Sometimes used as a functional term, where it is thought that all the cells in one column may serve the same sensory or motor function. Generally these functional columns are smaller than those shown in Figure 1.1(e).

Coronal: a plane perpendicular to the midline that passes vertically through the head.

Corpus callosum: a large bundle of nerve fibres that links the two cerebral hemispheres across the midline (see label 4 in Figure 1.1(b)).

Corticospinal tract: a large fibre bundle that passes mainly from the pre- and postcentral areas of the cerebral cortex (labelled 2 and 3 in Figure 1.1(a)) to the spinal cord. It plays a crucial role in the production of voluntary movements.

Cytoplasm: the contents of a cell, excluding the nucleus.

Cranial nerves: the 12 nerves that leave the brain and the brainstem on each side and that form a series, which continues as the spinal nerves.

Dendrite: one of several processes that that characterize the majority of nerve cells and that are readily distinguished from the single axon (see Figure 2.4).

Deep tegmental nucleus: a small group of cells in the midbrain (labelled 13 in Figure 6.1) that receives inputs from the mamillotegmental tract and contributes outputs that pass to the mamillary bodies through the mamillary peduncle. The functional roles of this well-studied group of cells still remain to be defined.

Dendritic spines: the very small protuberances that are present on many (but by no means all) dendrites, and that especially in the cerebral cortex can increase in size and number in relation to learning. Not shown in illustrations used in this book.

Dorsal: towards the back of the body. The developmental basis of the term makes it confusing in the brain, where the embryonic axis of the cerebrum lies at right angles to the embryonic axis of the brainstem in the adult. Consequently in the traditional naming of thalamic nuclei, the use of dorsal and ventral can be arbitrary and confusing.

Dorsal horn: the group of nerve cells that lie in the dorsal part of the spinal cord (see Figure 1.2).

Dorsal root: the nerve fibres that enter the dorsal part of the spinal cord and carry sensory inputs towards the thalamus.

Dorsal root ganglion: the nerve cells that give rise to the fibres of the dorsal root (labelled DRG in Figure 1.2).

Dorsal tegmental nucleus: a small group of cells that lies in the midbrain (labelled 9 in Figure 6.1). These cells integrate messages from the vestibular nuclei, which record movements of the head. The dorsal tegmental cells then in turn send messages about head direction to the lateral mamillary nucleus (see Figure 6.2) for relay first to the small anterodorsal thalamic nucleus and then to the retrosplenial cortex. These head direction cells contribute to the maps of the animal's space that are generated in the hippocampus. Compare to the deep tegmental nucleus, about whose functions we still know nothing.

Driver: when applied to the neural input to a thalamic or cortical cell, this refers to an input that carries a message for relay or processing, respectively.

Efference copies: copies of an instruction for an action. Sometimes called collateral discharges.

Efferent: an efferent message is one that is sent out to the motor apparatus for action.

Entorhinal cortex: a part of the hippocampal formation through which messages are passed from the cerebral cortex to the hippocampus itself.

Electron microscope: a microscope that uses beams of accelerated electrons rather than light waves. It provides much higher magnifications and reveals details not visible with light microscopes. Electron microscopes were available before the Second World War, but it was not until the early 1950s that methods of preparing neural tissues for study by electron microscopy were developed.

First-order thalamic nucleus: a thalamic nucleus that receives its driver inputs from subcortical centres.

Frontal lobe: the part of the cerebral hemispheres that lies anterior (in front of) the central sulcus (labelled F in Figure 1.1(a)).

GABA (GABAergic): abbreviation for gamma-aminobutyric acid, a neurotransmitter that generally in an adult mammal acts as an inhibitory transmitter, reducing the probability that the recipient cell will fire an action potential.

Gap junctions: these are junctions between two cells that allow the passage of small molecules from one cell to another, thus providing communication between the two cells, a possibility denied by the neuron doctrine.

Globus pallidus: one of the components of the basal ganglia (labelled 3 in Figure 1.1(d)).

Glomerulus: in the thalamus this term refers to an area of closely packed neural profiles making synaptic contacts with each other. The area is almost entirely free of glial processes and tends to be surrounded by sheets of glial cytoplasm. The term was first used in the olfactory bulb and cerebellum by light microscopists and was applied to the thalamus by Szentágothai in 1963.

Glutamatergic: this refers to cells and synapses whose neurotransmitter is glutamate.

Golgi method: a method developed by Camillo Golgi in 1873. It stained a small proportion of the nerve cells in a piece of tissue, revealing each cell completely by impregnating each with a mysterious 'black substance' (see Figure 1.1(e) on right, and Figures 2.4 and 2.6). The method was wonderfully revealing about the detailed nature of nerve cells, but the chemical basis of its success is not understood.

Grand mal epilepsy: a form of epilepsy that involves severe, involuntary movements of the body and limbs.

Gustatory pathways: neural pathways concerned with the sense of taste.

Gyrus: one of the convolutions of the cerebral cortex.

Head direction cells: cells recorded in the dorsal tegmental nucleus, lateral mamillary nucleus, the anterodorsal thalamic nucleus, and the retrosplenial cortex respond specifically to the direction in which the head is pointing. The messages that generate these responses come from the vestibular nuclei of the brainstem, are integrated in the dorsal tegmental nucleus, and sent to the lateral mamillary nucleus (see Figure 6.1).

Higher-order thalamic nucleus: a thalamic nucleus that receives many of its driver inputs from the cerebral cortex.

Hippocampus: a structure that lies deep in the temporal lobes of the brain and plays a role in the formation of new memories. It contains 'space cells', each closely linked to a particular portion of an animal's current space, which receive inputs from head direction cells.

Hypothalamus: a part of the brain that lies just below the thalamus and is importantly concerned with mechanisms that involve the control of the viscera. It was called the 'head ganglion of the visceral nervous system' by Sherrington (labelled 9 in Figure 1.1(b)).

Impulse: a nerve impulse is used as an alternative term for an action potential. *See also* action potential.

Inferior colliculus: a part of the midbrain in mammals that is concerned with hearing. It lies adjacent to the superior colliculus which is also concerned with vision and somatosensory pathways (see Figure 1.1(b) and label 13 in Figure 1.1(d)). The superior and inferior colliculi between them form the mammalian tectum (roof).

Internal capsule: a large structure made up of fibres passing between the cerebral cortex and the brainstem. Two parts, an anterior limb and a posterior limb, can be distinguished on horizontal sections of the brain in which the fibres are cut roughly perpendicularly (labelled 5 and 6 in Figure 1.1(d)).

Inferior olive: a group of cells providing inputs to the cerebellum.

Interneuron: a nerve cell whose axon has a local distribution, not extending significantly beyond the dendrites.

Initial segment: the slender part of an axon close to the nerve cell body, where the action potential originates (see Figure 2.4).

Insula: part of the cortex not seen on the surface of the brain because it is buried between the frontal and temporal lobes (labelled I in Figure 1.1(d)).

Ionotropic receptor: *see* receptor.

Kinaesthesis: (also kinaesthesia) the sensory modality concerned with the movements of limbs and joints.

Korsakoff's syndrome: a syndrome occurring in chronic alcoholics which includes severe memory losses and characteristically shows lesions of the mamillothalamic pathways at post-mortem.

Labelled lines: a view of individual neural pathways as each concerned with a particular identifiable function (e.g. nerve fibres that transmit information about painful stimuli do not transmit information about other functions, such as joint movements or pressure on the skin). The nerve cells receiving the information record the dynamics of the actual stimulus, but the bodily origin and the particular modality of the stimulus is a characteristic of each input fibre. This is sometimes expressed as Müller's law, formulated by Johannes Müller in 1826.

Lateral columns: the column of nerve fibres that lies between the dorsal and the ventral horns of the spinal cord. One of the important pathways in the lateral columns is the corticospinal tract.

Lateral geniculate nucleus: *see* thalamic nuclei.

Law of Bell and Magendie: a law that states that the sensory inputs (*see* afferent) from the body and the world reach the spinal cord and brainstem through the dorsal roots, whereas the motor outputs pass out from the spinal cord and brainstem through the ventral roots.

Law of dynamic polarization: a law proposed by Cajal and Van Gehuchten (see Chapter 5) stating that the inputs to a nerve cell travel along the dendrites and the outputs go through the axon. This was a wonderfully useful generalization about most nerve cells that conducted messages over long distances, but was not a general law about all nerve cells, some of which lack axons, while others, like dorsal root ganglion cells (see Figure 1.2) often lack dendrites and always receive their main inputs from an axon.

Lemniscus: *see* medial lemniscus.

Locomotor centre of the midbrain: a group of nerve cells in the midbrain that contributes to an animal's movements (see Chapter 1).

Magendie: *see* law of Bell and Magendie.

Mamillary bodies: two small, well-defined, spherical structures in the caudal part of the hypothalamus discussed in Chapter 6. They receive inputs from the midbrain and the fornix, and send outputs to the three anterior thalamic nuclei: anterodorsal, anteromedial, and anteroventral. We still remain ignorant about the messages that they send to the two largest anterior thalamic nuclei.

Mamillary peduncle: a small bundle of fibres that carries messages from the midbrain to the mamillary bodies (labelled 16 in Figure 6.1). Those that go to the lateral mamillary nucleus are messages about the direction of the head.

Mamillothalamic tract: a large, well-defined bundle that carries messages from the mamillary bodies to the three anterior thalamic nuclei (labelled 11 in Figure 6.1; see also Figure 6.4).

Mamillotegmental tract: a bundle of fibres that carries messages from the mamillary bodies to cell groups in the midbrain, one of which is concerned with the control of gaze (i.e. the direction in which the eyes are pointing) (labelled 12 in Figure 6.1; see also Figure 6.4).

Marchi method: a method for staining myelinated nerve fibres. The method played an important role in defining central neural pathways in the nineteenth and early twentieth centuries. Unmyelinated fibres (*see* myelin/myelin sheath), which include many nerve fibres in the hypothalamus and also the terminal segments of most nerve fibres, are not stained by this method.

Medial: towards the midline.

Medial dorsal nucleus: *see* thalamic nuclei.

Medial lemniscus: a sensory pathway that carries somatosensory messages (*see* somatosensory) from the caudal part of the brainstem to the thalamus, crossing the midline near its origin (see Figure 2.6).

Medulla: the most caudal part of the brainstem (labelled 6 in Figure 1.1(a)).

Medullary pyramid: *see* corticospinal tract; pyramid.

Membrane potential: the electric potential that exists across the surface membrane of a cell.

Mesencephalic motor centre: a cell group in the midbrain that contributes to the control of locomotion.

Metabotropic receptor: *see* receptor.

Midbrain: the part of the brainstem nearest to the cerebral hemispheres (labelled 12 in Figure 1.1(b)).

Mitochondria: small organelles found in most cells and commonly characterizing the presynaptic terminals of axons. They are responsible for the energy production in the region of the cell that they occupy.

Modulator: a neural input to a nerve cell that modulates the transmission of a message rather than carrying a message for transmission, as does a driver (*see* driver).

Motor cortex: the area of the cerebral cortex thought to be responsible for the initiation of voluntary movements. It occupies the precentral gyrus (labelled 2 in Figure 1.1(a)), where the several body parts are topographically represented. It gives rise to a large part, but by no means all, of the corticospinal tract.

Myelin/myelin sheath: a layer of lipid membranes that surrounds many axons, insulating the axons and increasing the speed at which the nerve fibre can conduct an action potential. The thickness of the myelin sheath increases with the diameter of the axon. The thinnest axons have no myelin sheath and conduct slowly.

Neural messages: the temporal patterns of action potentials that correspond to specific events in the body or the world.

Neurofibrils: these are fine fibrils that commonly characterize a nerve cell and its processes. They were revealed by light microscopists using methods based on the use of silver salts. Electron microscopic studies later showed them to be composed of many much finer processes, called neurofilaments. They are a part of a complex cytoplasmic skeleton largely concerned with the transport of material from one part of a nerve cell to another (see Figure 5.3).

Neurofilaments: *see* neurofibrils.

Neuron doctrine: a theory or doctrine that dominated studies of the nervous system for more than 75 years. It treated neurons as independent units in development, degeneration, regeneration, and function, and stressed that neurons never form fusions with each other.

It was opposed by the reticular theory. It is discussed in some detail in Chapter 5.

Neuronist: a supporter of the neuron doctrine.

Neurotransmitter: a chemical substance that is released by an active axon terminal and that can influence the membrane potential of the postsynaptic cell, either increasing the probability of a postsynaptic response (excitatory transmitters) or decreasing it (inhibitory transmitters).

Nigro-tectal pathway: an inhibitory pathway that passes from cells in a brainstem nucleus, the substantia nigra pars reticularis, to the superior colliculus. The pathway provides one important output of the basal ganglia.

Nissl method: a method of staining the cell bodies of nerve cells (see the two images on the left of Figure 1.1(e)).

Node of Ranvier: nodes that interrupt the myelin wrapping around individual nerve fibres. The interrupted myelin wrapping forces the membrane potential changes of the action potential to jump from one (uninsulated) node to the next. The resulting 'saltatory' conduction is significantly more rapid than that of an unmyelinated axon.

Nucleus: this has two meanings: (1) the nucleus of a cell which contains the DNA of the cell; (2) a reasonably well-defined grouping of cells in the central nervous system that often shares a set of connections and functions (e.g. the lateral geniculate nucleus of the thalamus or the dorsal tegmental nucleus of the midbrain).

Occipital lobe: the posterior part of the cerebral hemispheres (labelled O in Figure 1.1(a)).

Olfactory bulb: a pair of small club-like structures that receive inputs from the olfactory sensory receptors in the nose and send axons to the brain (labelled 1 in Figure 1.1(c)).

Olive: *see* inferior olive; superior olive.

Optic chiasm: a partial crossing the fibres that pass from the retina to the lateral geniculate nucleus and to the midbrain (labelled 3 in Figure 1.1(c); see Figure 8.1).

Optic nerve: the nerve that carries nerve fibres from the retina to the optic chiasm from which they pass to the lateral geniculate nucleus and the midbrain in the optic tract.

Papez circuit: the system of interconnected brain parts that includes the fibres of the postcommissural fornix going to the mamillary bodies, the mamillothalamic tract from the mamillary bodies to the anterior thalamic nuclei, the projections from the anterior thalamic nuclei to the cingulate cortex, the pathways from the cingulate cortex to the hippocampus, and the fornix fibres that leave the hippocampus, with some going to the mamillary bodies. See Papez (1937) and Chapter 6 for a fuller account.

Parasagittal: a plane of section that is parallel to the midline plane.

Parasympathetic system: the visceral nervous system (*see* visceral nervous system) has two parts that differ in their spinal origins, the distribution of their nerve cells, and their functions in ways that do not concern this book. One is the sympathetic system, the other the parasympathetic system.

Parietal cortex, parietal lobe: a part of the cerebral hemisphere that lies behind the central fissure and in front of the occipital lobe (labelled P in Figure 1.1(a)).

Parkinsonism: a clinical condition that produces motor deficits which gradually increase with age. They include tremors, rigidity of movements, and an unsteady gait, and relate to losses in the circuits of the basal ganglia.

Peduncle: a well-defined bundle of (usually myelinated) nerve fibres linked to a named part of the brain, for example, the cerebral peduncle, cerebellar peduncle, and mamillary peduncle.

Perception: a term used in this book to refer to a conscious sensory experience.

Perikaryon: the part of a cell that surrounds the nucleus.

Perineurium: a layer of connective tissue that surrounds bundles of nerve fibres

Peripheral nervous system: the parts of the nervous system that distribute messages between the central nervous system and the body.

Phagocytosis: the uptake of cellular and other debris in the body by a cell called a phagocyte.

Phrenologist: a person who studies the outlines and shape of the skull in order to interpret these in terms of the individual's cognitive and behavioural characteristics.

Pineal gland: a small gland that lies just above the back end of the thalamus at the midline. Descartes, looking for a midline structure, suggested it as the seat of the soul or the mind.

Pons: the part of the brainstem between the midbrain and the medulla (labelled 13 in Figure 1.1(b) and 6 in Figure 1.1(c)).

Pontine tegmental reticular nucleus: a group of cells close to the midline in the pons that plays a role in the control of gaze, that is, the position of the head and the eyes.

Postcentral gyrus: the postcentral sensory cortex or somatosensory cortex: the area of cortex (labelled 3 in Figure 1.1(a)) behind the central fissure (labelled 1 in Figure 1.1(a)). It receives topographically mapped sensory inputs from the body, matched across the central sulcus with a topographically organized motor map.

Postcommissural fornix: a part of a larger system of fornix fibres that leaves the hippocampus, arching below the corpus callosum and above the thalamus (labelled 7 in Figure 6.1). This fibre system splits at the anterior commissure, the precommissural fibres going to a region at the base of the frontal lobe in front of the hypothalamus (labelled 6 in Figure 6.1) and the postcommissural fibres going to the mamillary bodies and the anterior thalamus.

Posterior cingulate cortex: one of the three parts of the cingulate cortex (labelled 2 in Figure 1.1(b)). It receives inputs from the anteroventral thalamic nucleus.

Posterior column: the columns of fibres that lie between the two dorsal horns of the spinal cord, separated from each other by the midline.

Precentral gyrus: the cortical gyrus that lies just in front of the central sulcus (labelled 2 in Figure 1.1(a)). This area of cortex corresponds to the motor cortex (*see* motor cortex).

Precommissural fornix: *see* postcommissural fornix.

Pretectal nuclei: small cell groups that lie between the midbrain and the thalamus. Not shown in the figures. They play an important role in the control of the pupil and the lens of the eye and receive inputs from the eye and the cortex.

Putamen: one of the cell groups of the basal ganglia. Together with the caudate nucleus it forms a single functional unit called the striatum. The putamen and the caudate nucleus are separated from each other in the human brain by the anterior commissure (see labels 1, 2, and 5 in Figure 1.1(d)).

Pyramid: a swelling on the ventral surface of the medulla (labelled 8 in Figure 1.1(c)) which is formed by the pyramidal or corticospinal tract.

Pyramidal cell: these are cells that have a single apical dendrite and several basal dendrites. Several are shown in Figure 1.1(e). They are seen primarily in layers 2, 3, 5, and 6, with the largest in layer 5 (Cajal's layer 6 in Figure 1.1(e)), and these send their axons to subcortical regions or to the opposite hemisphere.

Pyramidal tract: *see* corticospinal tract.

Reafference: the change in sensory inputs produced by a movement of a sensory organ (e.g. the eye, the ear, or a finger).

Receptor: this term has two distinct meanings in the text. One describes sensory receptors, the cells that respond to stimuli from the body or the world—visual receptors in the retina, auditory receptors in the ear, tactile receptors in the skin, and so on. The other refers to complex molecules in the cell membrane that respond chemicals released by other cells. For nerve cells and for the text of this book, the most important such receptors lie in the postsynaptic cell membrane and play a role in synaptic transmission. Some, the ionotropic receptors, respond by directly opening channels in the membrane, whereas others, the metabotropic receptors, react through more complex and slower chain of chemical reactions. See Chapters 7 and 8.

Red nucleus: a cell group that plays an important role in the control of movements. It receives inputs from the motor cortex and from the cerebellum and sends outputs to the spinal cord.

Reduced silver methods: methods based on the use of silver nitrate solutions that demonstrated the structure of nerve cells primarily on the basis of the fibrillar contents (see Figures 5.3 and 7.2).

Relay: used here to describe the transfer of messages from one cell group to another, particularly for the relay of messages from the thalamus to the cerebral cortex. Hence, relay cells as the thalamic cells that transfer messages to cortex.

Reticular nucleus: *see* thalamic nuclei.

Reticular theory: a view of the nervous system as formed by a continuum of neural processes that are fused with each other. This had its roots in a holistic view of the brain's functions and was strongly opposed by the neuron doctrine or theory (*see* neuron doctrine and Chapter 5).

Reticularist: a supporter of the reticular theory.

Retinogeniculate axon: a nerve fibre that passes from the retina to the lateral geniculate nucleus of the thalamus.

Retina: the sheet of receptors and nerve cells at the back of the eye whose cells respond to light, and pass messages through complex neural interconnections within the retina itself, eventually to the lateral geniculate nucleus and to the superior colliculus and the pretectum in the midbrain.

Retrosplenial cortex: one of the three divisions of the cingulate cortex (labelled 3 in Figure 1.1(b)) that receives messages about head direction from the anterodorsal thalamic nucleus.

Romberg's test: a test for checking the action of the lower dorsal root axons in patients suffering from tabes dorsalis.

Rostral: towards the head.

Saccade: a rapid movement of the eyes.

Sagittal plane: the midline plane of the head and body.

Saphenous nerve: a peripheral nerve innervating the leg and the foot.

Sensory cortex: in general, an area of cortex receiving sensory inputs of any modality. Sometime applied specifically to the cortex of the postcentral gyrus (labelled 3 in Figure 1.1(a)), which is the somatosensory cortex.

Sensory pathways: any of the pathways that carry sensory messages to the brain or spinal cord or within the brain, including auditory, gustatory, olfactory, somatosensory, vestibular, and visual.

Sensory receptors: *see* receptor.

Somatosensory (or somatic sensory): this is a term used to refer to sensory pathways that are concerned with sensory stimuli coming from the limbs, the body, and the head; they include tactile, thermal, and pressure stimuli as well as painful stimuli and stimuli concerned with bodily movements. Hence, somatosensory pathways, and somatosensory cortex.

Spinal cord: the long slender neural structure that lies in the vertebral canal extending from the brainstem to the level of the second lumbar vertebra.

Spinal nerves: the nerves that form from the dorsal and ventral roots of the spinal cord (see Figure 1.2) and pass out on each side between the vertebrae.

Spine: *see* dendritic spines.

Spinothalamic pathway or tract: a sensory pathway transmitting information about painful and thermal stimuli which passes from spinal levels to the thalamus.

Startle response: a response, seen as a sudden movement of body and limbs, that can be elicited in an infant by a loud handclap and that serves to test the infant's hearing.

Striatum: a part of the basal ganglia that includes the caudate nucleus and the putamen (labelled 1 and 2 in Figure 1.1(d)).

Subiculum: a part of the hippocampal formation through which messages are passed from the cerebral cortex to the hippocampus itself.

Substantia nigra pars compacta: a group of cells that supplies a dopaminergic input to the striatum. The cells that synthesize the dopamine also make melanin in some species, producing the black appearance in human brains.

Substantia nigra pars reticularis: a group of cells adjacent to the pars compacta, that sends GABAergic, inhibitory inputs to several thalamic nuclei and to brainstem motor regions including the superior colliculus. See label SN in Figure 4.1, which shows the last of these pathways.

Sulcus: the furrow that separates two cortical gyri.

Superior cerebellar peduncle: *see* cerebellar peduncle.

Superior colliculus: the rostral of the two swellings forming the mammalian tectum (labelled SC in Figure 4.1, and labelled 12 in Figure 1.1(d)).

Superior olive: a group of cells in the brainstem serving as a relay in the auditory pathways.

Sympathetic system: *see* parasympathetic system.

Synapse: a one-way functional link (usually chemical) between two nerve cells. Electrical junctions which may act in both directions are an exception.

Synaptic bouton: the specialized part of an axon that forms the presynaptic part of a synaptic junction.

Synaptic transmitter: a chemical that when released from a presynaptic part of a synapse, the axon terminal, can act to change the local membrane potential of the postsynaptic cell.

Synaptic vesicle: small vesicles that contain synaptic transmitters.

Syncytium: a group of fused cells, sharing a continuous cytoplasm.

Tabes dorsalis: a clinical condition that produces degenerative changes in the dorsal root ganglion cells and also in their central and peripheral processes.

Tectum: the roof of the midbrain. In mammals it forms two swellings, the superior and inferior colliculi.

Tegmentum: the part of the midbrain that does not include the tectum or the cerebral peduncles.

Temporal lobe: the lobe of the cerebral hemisphere that lies below the frontal lobe (labelled T in Figure 1.1(a)).

Thalamic gate: the mechanisms of the thalamus that control the relay of messages to the cerebral cortex.

Thalamus: a large structure near the centre of the cerebral hemispheres that serves primarily as a relay station for sending messages from lower levels of the brain and spinal cord to the cerebral cortex.

Thalamic nuclei: the major cell groups of the thalamus defined by their inputs from different sources, their outputs to different cortical areas, and also by the grouping and the appearance of their cells. *Individual thalamic nuclei*—only thalamic nuclei that play a significant role in the text are included here:

Anterodorsal: the smallest of the three anterior thalamic nuclei, concerned with relaying messages about head direction (see Chapter 6). A nucleus that receives inputs from the lateral mamillary nucleus and relays messages to the retrosplenial cortex. A first-order nucleus.

Anteromedial: one of the three anterior thalamic nuclei whose functional properties remain to be defined. It relays messages from the mamillary bodies to the anterior cingulate cortex. A first-order nucleus.

Anteroventral: one of the three anterior thalamic nuclei whose functional properties remain to be defined. It relays messages from the mamillary bodies to the posterior cingulate cortex. A first-order nucleus.

Dorsal medial: a higher-order nucleus that relays messages to and receives messages from the frontal cortex.

Lateral geniculate: the thalamic nucleus that relays visual messages from the retina to the occipital lobe of the brain. A first-order nucleus.

Medial geniculate: this nucleus serves as a relay for auditory messages and has two major parts: a ventral part that provides a first-order relay and a dorsal part that is a higher-order relay.

Pulvinar: a higher-order nucleus that includes somatosensory and visual pathways.

Reticular nucleus: a slender group of nerve cells that lies immediately next to the lateral and ventral parts of the main part of the thalamus. These cells provide topographically organized modulatory, excitatory inputs from cortical layer 6 to specific parts of the thalamus, allowing the cortex to lower the thresholds of thalamic cells in a part of the thalamus functionally corresponding to a specific cortical area.

Ventral anterior: a higher-order nucleus linked to the frontal cortex.

Ventral lateral: a first-order nucleus that relays messages from the cerebellum to the motor cortex.

Ventral posterior lateral: a first-order nucleus that relays somatosensory messages from the body to the parietal cortex.

Ventral posterior medial: a first-order nucleus that relays somatosensory messages from the head to the parietal cortex.

Threshold: the level of a nerve cell's membrane potential just sufficient for it to fire an action potential.

Tonic mode of firing: regular frequency of the firing of action potentials in neurons; a pattern of firing in a thalamic relay cell that enables a message to be relayed in a form closely similar to that of the message received (see Chapter 8).

Tract: a neural pathway.

Triad: a pattern of synaptic contacts seen in the thalamus, in some other parts of the brain, and in the retina. It has three adjacent synaptic processes, one presynaptic to the other two, and of these two, one is also presynaptic to the other. In the thalamus the middle process is dendritic.

Transmitter: *see* neurotransmitter.

Ventral: towards the belly surface of the embryo. *See also* dorsal.

Ventral columns: the fibre columns of the spinal cord that lie between the two ventral horns of the grey matter.

Ventral horn: the group of nerve cells that lies in the ventral part of the spinal cord, within which lie the large motor cells that innervate the muscles (see Figure 1.2).

Ventral lateral nucleus: *see* thalamic nuclei.

Ventral posterior nucleus: *see* thalamic nuclei.

Ventral root: the fibre bundles that exit the ventral part of the spinal cord, carrying motor fibres (see Figure 1.2).

Ventricle: fluid-filled cavities that characterize each of the major parts of the brain appear in some of the figures in this book. The cerebral hemispheres each has a centrally placed, complexly curved lateral ventricle; the thalamus and hypothalamus lie on either side of a narrow midline third ventricle, the brainstem has a fourth ventricle that is joined to the third ventricle by the cerebral aqueduct, a narrow channel that separates the midbrain tectum from the tegmentum. It continues into the fourth ventricle, which relates to the pons and the medulla and, in turn, opens into a central canal of the spinal cord. The fluid that fills the ventricle is kept moving by a system of ciliary processes and plays an important role in the functions of the brain. However, the functions of this system play no role in this book other than their presence in illustrations.

Ventrobasal nucleus: a term sometimes used for ventral posterior nucleus (*see* thalamic nuclei).

Vestibular nuclei: a complex of nuclei in the brainstem that receive inputs from the vestibular nerves, send ascending pathways to brainstem structures and thalamus, and also send descending pathways to motor centres of the brainstem and spinal cord.

Vestibulo-ocular reflex: reflexes that link the movements of the eyes to the movements of the head so that we can keep our eyes fixated on an object as we move our heads.

Visceral functions: functions of the visceral organs, which include gut and gut derivatives like liver and pancreas, but also include kidneys, heart, blood vessels, lungs, sweat glands, and lachrymal glands.

Visceral nervous system: the part of the central and peripheral nervous systems that controls the visceral functions.

Visual cortex: the cortical area in the occipital lobes that receive visual inputs from the lateral geniculate nucleus (see Figure 6.1).

Weigert method: a method for staining myelinated nerve fibres.

Zona incerta: a group of cells lying ventral to the thalamus that receives input from the cerebral cortex and sends outputs to several brainstem centres and to the thalamus, where it is primarily inhibitory to its higher-order nuclei.

References

Adler J (1966). Chemotaxis in bacteria. *Science* **153**:708–16.

Adrian ED (1928). *The Basis of Sensation: The Action of the Sense Organs.* New York: WW Norton & Co.

Adrian ED (1947). General principles of nervous activity. *Brain* **70**:1–17.

Aggleton JP, O'Mara SM, Vann SD, Wright NF, Tsaniv M, Erichsen JT (2010). Hippocampal-anterior thalamic pathways for memory: uncovering a network of direct and indirect actions. *Eur J Neurosci* **31**:2292–307.

Akert K (Ed) (1981). *Biological Order and Brain Organization: Selected Works of W. R. Hess.* Berlin: Springer Verlag.

Akil H (2003). Scientific strategy in neuroscience: discovery science versus hypothesis-driven research. *Neurosci Q* Summer:4–5.

Andersen R (1995). Coordinate transformations and motor planning in posterior parietal cortex. In: Gazzaniga M (Ed), *The Cognitive Neurosciences*, pp 519–32. Cambridge, MA: MIT Press.

Apps MAJ, Ramnani N (2014). The anterior cingulate gyrus signals the net value of others' rewards. *J Neuroscience* **34**:6190–200.

Armstrong J, Richardson KC, Young JZ (1956). Staining neural end feet and mitochondria after postchroming and carbowax embedding. *Stain Technol* **31**:263–70.

Auerbach L (1898). Nervenendigungen in den Centralorganen. *Neurol Centralblatt* **17**:445–60.

Bach-y-Rita P (2004). Tactile sensory substitution studies. *Ann N Y Acad Sci* **1013**:83–91.

Baggini J (2015). *Freedom Regained: The Possibility of Free Will.* London: Granta Publications.

Bajo VM, Nodal FR, Bizley JK, Moore DR, King AJ (2007). The ferret auditory cortex: descending projections to the inferior colliculus. *Cereb Cortex* **17**:475–91.

Bard P, Mountcastle VB (1948). Some forebrain mechanisms involved in expression of rage with special reference to suppression of angry behavior. *Res Publ Assoc Res Nerv Ment Dis* **27**:362–404.

Bartels A, Logothetis NK, Moutoussis K (2008). fMRI and its interpretations: an illustration on directional selectivity in area V5/MT. *Trends Neurosci* **31**:444–53.

Beauchamp TL (Ed) (1999). *David Hume: An Enquiry Concerning Human Understanding.* Oxford: Oxford University Press.

Bell C (1811). *Idea of a New Anatomy of the Brain; Submitted for the Observations of His Friends.* London: Strahan and Preston. Reprinted in *Medical Classics* 1936; **1**:105–20.

Bender DB (1983). Visual activation in the primate pulvinar depends on cortex but not colliculus. *Brain Res* 279:258–61.

Bennett MV, Aljure E, Nakajima Y, Pappas GD (1963). Electrotonic junctions between teleost spinal neurons: electrophysiology and ultrastructure. *Science* 141:262–4.

Berg HC (1975). Bacterial behavior. *Nature* 254:389–92.

Berlucchi G (1999). Some aspects of the history of the law of dynamic polarization of the neuron. From William James to Sherrington, from Cajal and Van Gehuchten to Golgi. *J Hist Neurosci* 8:191–201.

Bernard C (1974). *Lectures on the Phenomena Common to Animals and Plants* (Hoff HE, Guillemin R, Guillemin L, Trans). Springfield, IL: Charles C Thomas.

Bickford ME, Zhou N, Krahe TE, Govindaiah G, Guido W (2015). Retinal and tectal 'driver-like' inputs converge in the shell of the mouse dorsal lateral geniculate nucleus. *J Neurosci* 35:10523–34.

Billig M (2013). *Learn to Write Badly: How to Succeed in the Social Sciences.* Cambridge: Cambridge University Press.

Blakemore SJ, Wolpert DM, Frith CD (1998). Central cancellation of self-produced tickle sensation. *Nature Neurosci* 1:635–40.

Blakemore SJ, Wolpert DM, Frith CD (2002). Abnormalities in the awareness of action *Trends Cogn Sci* 6:237–42.

Bompas A, O'Regan JK (2006). Evidence for a role of action in colour perception. *Perception* 35:65–78.

Böszörményi L (1997). Informatik in der (Waldorf) Schule. *Erziehungskunst* 2:1–8.

Bourassa J, Pinault D, Deschênes M (1995). Corticothalamic projections from the cortical barrel field to the somatosensory thalamus in rats: a single fiber study using biocytin as an anterograde tracer. *Eur J Neurosci* 7:19–30.

Boycott BB, Gray, EG, Guillery RW (1961). Synaptic structure and its alteration with environmental temperature: a study by light and electron microscopy of the central nervous system of lizards. *Proc Roy Soc B* 154:151–72.

Broca P (1863). Localisation des functions cérébrales. Siège du langage articulé. *Bull Soc Anthropol Paris* 4:200–3.

Brodal A (1981). *Neurological Anatomy in Relation to Clinical Medicine*, 3rd ed. New York: Oxford University Press.

Bronson GW (1990). Changes in infants' visual scanning across the 2- to 14-week age period. *J Exp Child Psychol* 49:101–125.

Brown-Séquard CE (1868). Nouvelles recherches sur le trajet des diverse espèces de conducteurs d'impressions sensitives dans la moelle épiniere. *Arch Physiol Normale et Pathol* 1:610–21, 716–24.

Brown-Séquard CE (1869). Nouvelles recherches sur le trajet des diverse espècesde conducteurs d'impressions sensitives dans la moelle épiniere (troisième partie). *Arch Physiol Normale et Pathol* 2:236–50.

Bullock TH, Bennett MVL, Johnston D, Josephson R, Marder E, Fields RD (2005). The neuron doctrine redux. *Science* 310:791–3.

Buneo CA, Andersen RA (2006). The posterior parietal cortex: sensorimotor interface for the planning and online control of visually guided movements. *Neuropsychologia* 44:2594–606.

Campbell AW (1905). *Histological Studies of the Localisation of Cerebral Function.* Cambridge: Cambridge University Press.

Cerkevich CM, Lyon DC, Balaram P, Kaas JH (2014). Distribution of cortical neurons projecting to the superior colliculus in macaque monkeys. *Eye Brain* 23:121–37.

Chalmers DJ (2013). How can we construct a science of consciousness? *Ann N Y Acad Sci* 1303:25–35.

Chalmers DJ, Jackson F (2001). Conceptual analysis and reductive explanation. *Philos Rev* 110:315–61.

Chalupa LM (1991). Visual function of the pulvinar. In: Leventhal AG (Ed), *The Neural Basis of Visual Function*, pp 140–59. New York: Macmillan Press.

Chalupa LM, Abramson BP (1988). Receptive field properties in the tecto- and striate-recipient zones of the cat's lateral posterior nucleus. *Prog Brain Res* 75:85–94.

Churchland PS (2002). *Brain-Wise: Studies in Neurophilosophy.* Cambridge, MA: MIT Press.

Churchland PS, Sejnowski TJ (1992). *The Computational Brain.* Cambridge, MA: MIT Press.

Churchland PS, Ramachandran VS, Sejnowski TJ (1994). A critique of pure vision. In: Koch C, Davis JL (Eds), *Large-Scale Neuronal Theories in the Brain*, pp 23–65. Cambridge, MA: MIT Press.

Clark A (1998). *Being There: Putting Brain Body and World Together Again.* Cambridge, MA: MIT Press.

Clark A (2008). *Supersizing the Mind: Embodiment, Action and Cognitive Extension.* Oxford: Oxford University Press.

Colby CL, Duhamel JR, Goldberg ME (1996). Visual presaccadic, and cognitive activation of single neurons in monkey lateral intraparietal cortex. *J Neurophsyiol* 76:2841–52.

Colonnier M, Guillery RW (1964). Synaptic organization in the lateral geniculate nucleus of the monkey. *Z Zellforsch* 62:333–55.

Conant JB (1947). On understanding science. *Am Sci* 35:33–55.

Cox CL, Denk W, Tank DW, Svoboda K (2000). Action potentials reliably invade axonal arbors of rat neocortical neurons. *Proc Natl Acad Sci USA* 97:9724–8.

Crabtree JW (1992). The somatotopic organization within the rabbit's thalamic reticular nucleus. *Eur J Neurosci* 4:1343–51.

Crabtree JW (1996). Organization of the somatosensory sector of the cat's thalamic reticular nucleus. *J Comp Neurol* 366:207–22.

Crabtree JW (1998). Organization in the auditory sector of the cat's thalamic reticular nucleus. *J Comp Neurol* **390**:167–82.

Crick F (1984). Function of the thalamic reticular complex: the searchlight hypothesis. *Proc Natl Acad Sci USA* 1984; **81**:4586–90.

Crick FC, Koch C (2005). What is the function of the claustrum? *Philos Trans R Soc Lond B Biol Sci* **360**:1271–9.

Cruce JA (1977). An autoradiographic study of the descending connections of the mammillary nuclei of the rat. *J Comp Neurol* **176**:631–44.

Dai C, Fridman GY, Davidovics NS, Chiang B, Ahn JH, Della Santina CC (2011). Restoration of 3D vestibular sensation in rhesus monkeys using a multichannel vestibular prosthesis. *Hear Res* **281**:74–83.

Damasio A (2006). *Descartes' Error*. London: Vintage.

Danilov YP, Tyler ME, Skinner KL, Hogle RA, Bach-Y-Rita P (2007). Efficacy of electrotactile vestibular substitution in patients with peripheral and central vestibular loss. *J Vestib Res* **17**:119–30.

Deeg KE, Aizenman CD (2011). Sensory modality-specific homeostatic plasticity in the developing optic tectum. *Nat Neurosci* **14**:548–50.

Dehaene S (2014). *Consciousness and the Brain*. New York: Viking.

Deiters O (1865). *Untersuchungen über Gehirn und Rückenmark des Menschen und der Säugethiere*. Braunschweig: Friedrich Vieweg und Sohn.

Deiters VS, Guillery RW (2013). Otto Friedrich Karl Deiters (1834–1863). *J Comp Neurol* **521**:1929–53.

De Robertis E (1959). Submicroscopic morphology of the synapse. *Internatl Rev Cytol* **8**:61–96.

De Robertis E, Bennett HS (1955). Some features of the submicroscopic morphology of synapses in frog and earthworm. *J Biophys Biochem Cytol* **1**:47–58.

Descartes R (1662). *De Homine*. [See **Finger S** (1994). *Origins of Neuroscience: A History of Explorations into Brain Functions*. Oxford: Oxford University Press.]

Deschênes M, Bourassa J, Pinault D (1994). Corticothalamic projections from layer V cells in rat are collaterals of long-range corticofugal axons. *Brain Res* **664**:215–19.

Diamond ME Armstrong-James M, Budway MJ, Ebner EF (1992). Somatic sensory responses in the rostral sector of the posterior group (POm) and in the ventral posterior medial nucleus (VPM) of the rat thalamus: dependence on the barrel field cortex. *J Comp Neurol* **319**:66–84.

Dillingham CM, Frizzati A, Nelson AJ, Vann SD (2015). How do mammillary body inputs contribute to anterior thalamic function? *Neurosci Biobehav Rev* **54**:108–19.

Dobson V, Teller DY (1978). Visual acuity in human infants: a review and comparison of behavioral and electrophysiological studies. *Vision Res* **18**:1469–83.

Domesick VB (1969). Projections from the cingulate cortex in the rat. *Brain Res* **12**:296–320.

Dräger UC, Hubel DH (1975). Responses to visual stimulation and relationship between visual, auditory, and somatosensory inputs in mouse superior colliculus. *J Neurophysiol* **38**:690–713.

Dumont JR, Taube JS (2015). The neural correlates of navigation beyond the hippocampus. *Prog Brain Res* **219**:83–102.

Eccles J, Gibson W (1979). *Sherrington: His Life and Thought*. Berlin: Springer International.

Economides M, Guitart-Masip M, Kurth-Nelson Z, Dolan RJ (2014). Anterior cingulate cortex instigates adaptive switches in choice by integrating immediate and delayed components of value in ventromedial prefrontal cortex. *J Neurosci* **34**:3340–9.

Elliott-Smith G (1910). Some problems relating to the evolution of the brain. *Lancet* **1**:1–8:147–53, 221–7.

Erişir A, Van Horn SC, Sherman SM (1997). Relative numbers of cortical and brainstem inputs to the lateral geniculate nucleus. *Proc Natl Acad Sci USA* **94**:517–20.

Etkin A, Egner T, Kalisch R (2011). Emotional processing in anterior cingulate and medial prefrontal cortex. *Trends Cogn Sci* **15**:85–93.

Evarts EV, Tanji J (1976). Reflex and intended responses in motor cortex pyramidal tract neurons of monkey. *J Neurophysiol* **39**:1069–80.

Fatt B, Katz B (1951). An analysis of the end-plate potential recorded with an intracellular electrode. *J Physiol* **115**:320–37.

Fatt P, Katz B (1952). Spontaneous subthreshold activity at motor nerve endings. *J Physiol* **117**:109–28.

Feinberg TE, Mallatt JM (2016). *The Ancient Origins of Consciousness: How the Brain Created Experience*. Cambridge, MA: MIT Press.

Felleman DJ, Van Essen DC (1991). Distributed hierarchical processing in the primate cerebral cortex. *Cerebral Cortex* **1**:1–47.

Finger S (1994). *Origins of Neuroscience: A History of Explorations into Brain Functions*. New York: Oxford University Press.

Finkelstein A, Derdikman D, Rubin A, Foerster JN, Las L, Ulanovsky N (2015). Three-dimensional head-direction coding in the bat brain. *Nature* **517**:159–64.

Firestein S (2012). *Ignorance: How it Drives Science*. Oxford: Oxford University Press.

Fornito A, Yücel M, Wood SJ, Bechdolf A, Carter S, Adamson C, Velakoulis D, Saling MM, McGorry PD, Pantelis C (2009). Anterior cingulate cortex abnormalities associated with a first psychotic episode in bipolar disorder. *Br J Psychiatry* **194**:426–33.

Foster DH (2011). Color constancy. *Vision Res* **51**:674–700.

Fritsch G, Hitzig E (1870). Über die elektrische Erregbarkeit des Grosshirns. *Arch Anat Physiol* **37**:300–32.

Furshpan EJ, Potter DD (1959). Transmission at the giant motor synapses of the crayfish. *J Physiol* **145**:289–325

Georgopoulos AP, Kalaska JF, Caminiti R, Massey JT (1982). On the relations between the directions of two-dimensional arm movements and cell discharge in primate motor cortex *J Neurosci* 2:1527–37.

Gibson JJ (1986). *The Ecological Approach to Visual Perception*. Hillsdale, NJ: Lawrence Erlbaum Associates.

Glees P, Le Gros Clark WE (1941). The termination of optic fibres in the lateral geniculate body of the monkey. *J Anat* 75:295–308.

Goense J, Merkle H, Logothetis NK (2012). High-resolution fMRI reveals laminar differences in neurovascular coupling between positive and negative BOLD responses. *Neuron* 76:629–39.

Gopnik A (2009). Could David Hume have known about Buddhism? Charles Francois Dolu, the Royal College of La Flèche, and the Global Jesuit Intellectual Network. *Hume Stud* 35:5–28.

Gray EG (1959). Axo-somatic and axo-dendritic synapses of the cerebral cortex: an electron microscope study. *J Anat* 93:420–33.

Gray EG, Guillery RW (1966). Synaptic morphology in the normal and degenerating nervous system. *Int Rev Cytol* 19:111–82.

Grillner S (2003). The motor infrastructure: from ion channels to neuronal networks. *Nat Rev Neurosci* 4:573–86.

Grillner S (2015). Action: the role of motor cortex challenged. *Curr Biol* 25:R508–R511.

Grillner S, Robertson B (2016). The basal ganglia over 500 million years. *Curr Biol* 26:R1088–R1100.

Grillner S, Kozlov A, Dario P, Stefanini C, Menciassi A, Lansner A, Hellgren KJ (2007). Modeling a vertebrate motor system: pattern generation, steering and control of body orientation. *Prog Brain Res* 165:221–34.

Grünbaum AFS, Sherrington CS (1901). Observations on the physiology of the cerebral cortex in anthropoid apes. *Proc R Soc Lond* 69:206–9.

Guillery RW (1954). *The Mammillary Bodies and Their Connections*. PhD Thesis, University College London.

Guillery RW (1956). Degeneration in the post-commissural fornix and the mamillary peduncle of the rat. *J Anat* 90:350–70.

Guillery RW (1957). Degeneration in the hypothalamic connexions of the albino rat. *J Anat* 91:91–115.

Guillery RW (1995). Anatomical evidence concerning the role of the thalamus in corticocortical communication: a brief review. *J Anat* 187:583–92.

Guillery RW (2005). Observations of synaptic structures: origins of the neuron doctrine and its current status. *Phil Trans R Soc B* 360:1281–307.

Guillery RW (2007). Relating the neuron doctrine to the cell theory. Should contemporary knowledge change our view of the neuron doctrine? *Brain Res Rev* 55:411–21.

Guillery RW, Feig SL, Van Lieshout DP (2001). Connections of higher order relays in the thalamus: a study of corticothalamic connections in cats. *J Comp Neurol* 438:66–85.

Gunning FM, Cheng J, Murphy CF, Kanellopoulos D, Acuna J, Hoptman MJ, Klimstra S, Morimoto S, Weinberg J, Alexopoulos GS (2009). Anterior cingulate cortical volumes and treatment of geriatric depression. *Int J Geriatr Psychiatry* 24:829–36.

Hall NJ, Colby CL (2011). Remapping for visual stability. *Phil Trans R Soc B* 366:528–39.

Hámori J, Pasik T, Pasik P, Szentágothai J (1974). Triadic synaptic arrangements and their possible significance in the lateral geniculate nucleus of the monkey. *Brain Res* 80:379–93.

Harding BN (1973). An ultrastructural study of the termination of afferent fibres within the ventrolateral and centre median nuclei of the monkey thalamus. *Brain Res* 54:341–6.

Harris JA (2015). *Hume: An Intellectual Biography.* Cambridge: Cambridge University Press.

Harrison JM, Warr WB (1962). A study of the cochlear nuclei and ascending auditory pathways of the medulla. *J Comp Neurol* 119:341–79.

Harting JK, Updyke BV, Van Lieshout DP (1992). Corticotectal projections in the cat: anterograde transport studies of twenty-five cortical areas. *J Comp Neurol* 324:379–414.

Hayes DJ, Northoff G (2012). Common brain activations for painful and non-painful aversive stimuli. *BMC Neurosci* 13:13–60.

Held H (1906). Zur weiteren Kenntnis der Nervenendfüsse und zur strukur der Sehzellen. *Abhand der Kgl Sächs Gesellsch d Wissenschft math-phys KL* 29:143–52.

Helmholtz H von (1868). Recent progress in the theory of vision. In: Kahl R (Ed and Trans) (1971). *Selected Writings of Hermann von Helmholtz*, pp 144–222. Middletown, CT: Wesleyan University Press.

Helmholtz H von (1873). The recent progress of the theory of vision. In: Atkinson E (Ed), *Popular Lectures on Scientific Subjects*, pp 197–316. New York: D. Appleton and Company.

Herrick CJ (1948). *The Brain of the Tiger Salamander, Ambystoma tigrinum.* Chicago, IL: University of Chicago Press.

Hess BJ, Blanks RH, Lannou J, Precht W (1989). Effects of kainic acid lesions of the nucleus reticularis tegmenti pontis on fast and slow phases of vestibulo-ocular and optokinetic reflexes in the pigmented rat. *Exp Brain Res* 74:63–79.

Hess R Jr, Akert K, Koella W (1950). Bioelectric potentials of the cortex and thalamus, and their modification by stimulation of the hypnic center in cats. *Rev Neurol (Paris)* 83:537–44.

Hilbert D (2005). Color constancy and the complexity of color. *Philos Top* 33:141–58.

Hodgkin AL, Huxley AF (1952a). Currents carried by sodium and potassium ions through the membrane of the giant axon of Loligo. *J Physiol* 116:449–72.

Hodgkin AL, Huxley AF (1952b). The components of membrane conductance in the giant axon of Loligo. *J Physiol* 116:473–96.

Hoff EC, Hoff HE (1934). Spinal terminations of the projection fibres from the motor cortex of primates. *Brain* 57:454–74.

Holroyd CB, Yeung N (2012). Motivation of extended behaviors by anterior cingulate cortex. *Trends Cogn Sci* 2012:122–8.

Holstege G, Blok BF, Ralston DD (1988). Anatomical evidence for red nucleus projections to motoneuronal cell groups in the spinal cord of the monkey. *Neurosci Lett* 95:97–101.

Holstege JC, Kuypers HG (1987). Brainstem projections to spinal motoneurons: an update. *Neuroscience* 23:809–21.

Hubel DH (2005). David H. Hubel. In: Hubel DH, Wiesel TN, *Brain and Visual Perception: The Story of a 25-year Collaboration*, pp 5–24. Oxford: Oxford University Press.

Hubel DH, Wiesel TN (1961). Integrative action in the cat's lateral geniculate body *J Physiol* 155:385–98.

Hubel DH, Wiesel TN (1962). Receptive fields, binocular interaction and functional architecture in the cat's visual cortex. *J Physiol* 160:106–54.

Huerta MF, Harting JK (1982). Tectal control of spinal cord activity: neuroanatomical demonstration of pathways connecting the superior colliculus with the cervical spinal cord grey. *Prog Brain Res* 57:293–328.

Hughlings Jackson J (1881). Remarks on dissolution of the nervous system as exemplified by certain post-epileptic conditions. *Med Press Circular* 1:329–32.

Hume D (1739). *A Treatise of Human Nature*. Project Gutenberg, 2010 [EBook #4705]. Choat C, Widger D (Prod.)

Hume D (1748). *An Enquiry Concerning Human Understanding*. [See Beauchamp TL (Ed) (1999). *David Hume: An Enquiry Concerning Human Understanding*. Oxford: Oxford University Press.]

Humphrey NK, Weiskrantz L (1967). Vision in monkeys after removal of the striate cortex. *Nature* 215:595–7.

Hurley SL (1998). *Consciousness in Action*. Cambridge, MA: Harvard University Press.

Hurley S (2001). Perception and action: alternative views. *Synthese* 129:3–40.

Janušonis S (2015). Book review: a deep look at the thalamocortical continuum. *Vis Neurosci* 32:e024.

Jelbert SA, Taylor AH, Gray RD (2015). Reasoning by exclusion in New Caledonian crows (Corvus moneduloides) cannot be explained by avoidance of empty containers. *J Comp Psychol* 129:283–90.

Jones EG (2007). *The Thalamus*. Cambridge: Cambridge University Press.

Jones EG, Powell TP (1969a). An electron microscopic study of the mode of termination of cortico-thalamic fibres within the sensory relay nuclei of the thalamus. *Proc R Soc Lond B Biol Sci* **172**:173–85.

Jones EG, Powell TP (1969b). Electron microscopy of synaptic glomeruli in the thalamic relay nuclei of the cat. *Proc R Soc Lond B Biol Sci* **172**:153–71.

Jones EG, Powell TPS (1970). An anatomical study of converging sensory pathways within the cerebral cortex of the monkey. *Brain* **93**:793–820.

Jones EG, Rockel AJ (1971). The synaptic organization of the medial geniculate body and afferent fibres arising from the inferior colliculus. *Z Zellforsch* 113:44–66.

Kaas JH, Guillery RW, Allman JM (1972). Some principles of organization in the dorsal lateral geniculate nucleus. *Brain Behav Evol* **6**:253–99.

Kaas JH, Merzenich MM, Killackey HP (1983). The reorganization of somatosensory cortex following peripheral nerve damage in adult and developing mammals. *Annu Rev Neurosci* **6**:325–56.

Kacelnik O, Nodal FR, Parsons CH, King AJ (2006). Training-induced plasticity of auditory localization in adult mammals. *PLoS Biol* **4**:e71.

Kaiser J (2014). NIH puts massive U.S. children's study on hold. *Science* **344**:1327.

Kandel ER, Schwartz JH, Jessell TM (1991). *Principles of Neuroscience*, 3rd ed. New York: Elsevier.

Kardamakis AA, Saitoh K, Grillner S (2015). Tectal microcircuit generating visual selection commands on gaze-controlling neurons. *Proc Natl Acad Sci USA* 112:E1956–65.

Kass JH (2011). Reconstructing the areal organization of the neocortex of the first mammals. *Brain Behav Evol* **78**:7–21.

Kawai R, Markman T, Poddar R, Ko R, Fantana A, Dhawale AK, Kampff AR, Ölveczky B (2015). Motor cortex is required for learning but not for executing a motor skill. *Neuron* **86**:800–12.

Keliris GA, Logothetis NK, Tolias AS (2010). The role of the primary visual cortex in perceptual suppression of salient visual stimuli. *J Neurosci* **30**:12353–65.

Kenny A (1989). *The Metaphysics of Mind.* Oxford: Oxford University Press.

Kidd M (1962). Electron microscopy of the inner plexiform layer of the retina in the cat and pigeon. *J Anat* **96**:179–87.

Killackey HP, Sherman SM (2003). Corticothalamic projections from the rat primary somatosensory cortex. *J Neurosci* **23**:7381–4.

Kölliker A (1896). *Handbuch der Gewebelehre des Menschen*, 6th ed, vol 2. Leipzig: W Engelmann.

Kolling N, Behrens TE, Mars RB, Rushworth MF (2012). Neural mechanisms of foraging. *Science* **336**:95–8.

Korsakoff SS (1887). Disturbances of psychic functions in alcoholic paralysis and its relation to the disturbance of the psychic sphere in multiple neuritis of non-alcoholic origin. *Vestnik Psichiatrii* **4**:2.

Kuhn TS (1962). *The Structure of Scientific Revolutions*. Chicago, IL: University of Chicago Press.

Land EH (1974). The retinex theory. *Sci Am* 52:247–64.

Landgren S, Phillips CG, Porter R (1962). Minimal synaptic actions of pyramidal impulses on some alpha motoneurones of the baboon's hand and forearm. *J Physiol* 161:91–111.

Larkum ME, Zhu JJ, Sakmann B (2001). Dendritic mechanisms underlying the coupling of the dendritic with the axonal action potential initiation zone of adult rat layer 5 pyramidal neurons. *J Physiol* 533:447–66.

Laurens J, Kim B, Dickman JD, Angelaki DE (2016). Gravity orientation tuning in macaque anterior thalamus. *Nat Neurosci* 19:1566–8.

Lavallée DG, Urbain N. Dufresne C, Bokor H, Acsády L, Deschênes M (2005). Feedforward inhibitory control of sensory information in higher-order thalamic nuclei. *J Neurosci* 25:7489–98.

Lawrence DG, and Kuypers HG (1968a). The functional organization of the motor system in the monkey. I. The effects of bilateral pyramidal lesions. *Brain* 91:1–14.

Lawrence DG, Kuypers HG (1968b). The functional organization of the motor system in the monkey. II. The effects of lesions of the descending brain-stem pathways. *Brain* 91:15–36.

Lee AM, Hoy JL, Bonci A, Wilbrecht L, Stryker MP, Niell CM (2014). Identification of a brainstem circuit regulating visual cortical state in parallel with locomotion. *Neuron* 83:455–66.

Lee Y, Lopez DE, Meloni EG, Davis M (1996). Primary acoustic startle pathway: obligatory role of cochlear root neurons and the nucleus reticularis pontis caudalis. *J Neurosci* 16:3775–89.

Leopold DA, Logothetis NK (1996). Activity changes in early visual cortex reflect monkeys' percepts during binocular rivalry. *Nature* 379:549–53.

Loewy AD, Spyer KM (1990). *Central Regulation of Autonomic Functions*. New York: Oxford University Press.

Logan CJ, Breen AJ, Taylor AH, Gray RD, Hoppitt WJ (2016). How New Caledonian crows solve novel foraging problems and what it means for cumulative culture. *Learn Behav* 44:18–28.

Logothetis N (1998). Object vision and visual awareness. *Curr Opin Neurobiol* 8:536–44.

Logothetis NK (2002). The neural basis of the blood-oxygen-level dependent functional magnetic resonance imaging signal. *Philos Trans R Soc Lond B Biol Sci* 357:1003–37.

MacKay DM (1956). The epistemological problem for automomata. In: Shannon CE, McCarthy J (Eds), *Automoata Studies*, pp 235–52. Princeton, NJ: Princeton University Press.

MacKay DM (1982). Ourselves and our brains: duality without dualism. *Psychoneuroendocrinology* 7:285–94.

Magendie F (1822). Experiences sur les fonctiones des racines des nerfs rachidiens. *J Physiol Exp Pathol* 2:276–9.

Majarossy K, Kiss A (1976). Specific patterns of neuron arrangement and of synaptic articulation in the medial geniculate body. *Exp Brain Res* 26:1–17.

Mason A, Ilinsky IA, Beck S, Kultas-Ilinsky K (1996). Reevaluation of synaptic relationships of cerebellar terminals in the ventral lateral nucleus of the rhesus monkey thalamus based on serial section analysis and three-dimensional reconstruction. *Exp Brain Res* 109:219–39.

Mathers LH (1972). The synaptic organization of the cortical projection to the pulvinar of the squirrel monkey. *J Comp Neurol* 146:43–60.

Maximow AA (1938). *A Textbook of Histology*, 3rd ed. Philadelphia, PA: WB Saunders.

Mease RA, Metz M, Groh A (2016). Cortical sensory responses are enhanced by the higher-order thalamus. *Cell Rep* 14:208–15.

Medawar PB (1969). *Induction and Intuition in Scientific Thought*. London: Methuen & Co.

Melcher D, Colby CL (2008). Trans-saccadic perception. *Trends Cogn Sci* 12:466–73.

Méndez-López M, Méndez M, López L, Arias JL (2009). Spatial working memory learning in young male and female rats: involvement of different limbic system regions revealed by cytochrome oxidase activity. *Neurosci Res* 65:28–34

Merfeld DM, Gong W, Morrissey J, Saginaw M, Haburcakova C, Lewis RF (2006). Acclimation to chronic constant-rate peripheral stimulation provided by a vestibular prosthesis. *IEEE Trans Biomed Eng* 53:2362–72.

Merleau-Ponty M (1958). *Phenomenology of Perception* (Smith C, Trans). London: Routledge Classics.

Miller GA, Galanter E, Pribram, KH (1960). *Plans and the Structure of Behavior*. New York: Holt, Rinehart & Winston.

Milner B (1958). Psychological defects produced by temporal lobe excision. *Res Publ Assoc Res Nerv Ment Dis* 36:244–57.

Montero VM, Guillery RW, Woolsey CN (1977). Retinotopic organization within the thalamic reticular nucleus demonstrated by a double label autoradiographic technique. *Brain Res* 138:407–21.

Mossio M, Taraborelli D (2008). Action-dependent perceptual invariants: from ecological to sensorimotor approaches. *Conscious Cogn* 17:1324–40.

Mountcastle VB, Henneman E (1952). The representation of tactile sensibility in the thalamus of the monkey. *J Comp Neurol* 97:409–39.

Müller J (1826). *Zur vergleichenden Physiologie des Gesichtsinnes des Menschen und der Thiere*. Leipzig: C Cnobloch.

Müller-Preuss P, Jürgens U (1976). Projections from the 'cingular' vocalization area in the squirrel monkey. *Brain Res* 103:29–43.

Nicholls JG (2007). How acetylcholine gives rise to current at the endplate. *J Physiol* 578:621–2.

Nodal FR, Kacelnik O, Bajo VM, Bizley JK, Moore DR, King AJ (2010). Lesions of the auditory cortex impair azimuthal sound localization and its recalibration in ferrets. *J Neurophysiol* **103**:1209–25.

Noë A (2004). *Action in Perception*. Cambridge, MA: MIT Press.

Nurse P (2015). Trust in science. *Notes Rec R Soc* **69**:17–222.

Ocaña FM, Suryanarayana SM, Saitoh K, Kardamakis AA, Capantini L, Robertson B, Grillner S. The lamprey pallium provides a blueprint of the mammalian motor projections from cortex. *Curr Biol* **25**:413–23.

Ogren MP, Hendrickson AE (1979). The morphology and distribution of striate cortex terminals in the inferior and lateral subdivisions of the Macaca monkey pulvinar. *J Comp Neurol* **188**:179–99.

O'Keefe J, Dostrovsky J (1971). The hippocampus as a spatial map. Preliminary evidence from unit activity in the freely-moving rat. *Brain Res* **34**:171–5.

O'Keefe J (1978). *The Hippocampus as a Cognitive Map*. Oxford: Clarendon Press.

O'Regan JK, Noë A (2001). A sensorimotor account of vision and visual consciousness. *Behav Brain Sci* **24**:939–73.

Palay SL, Palade GE (1955). The fine structure of neurons. *J Biophys Biochem Cytol* **1**:69–88.

Papez J (1937). A proposed mechanism of emotion. *Arch Neurol Psychiat* **38**:725–43.

Pasik T, Pasik P, Hamori J, Szentagothai J (1973). 'Triadic' synapses and other articulations of interneurons in the lateral geniculate nucleus of rhesus monkeys. *Trans Am Neurol Assoc* **98**:293–5.

Passingham RE, Bengtsson SL, Lau HC (2010). Medial frontal cortex: from self-generated action to reflection on one's own performance. *Trends Cogn Sci* **14**:16–21.

Penfield W, Boldrey E (1937). Somatic motor and sensory representation in the cerebral cortex of man as studied by electrical stimulation. *Brain* **60**:389–443.

Petrof I, Sherman SM (2009). Synaptic properties of the mammillary and cortical afferents to the anterodorsal thalamic nucleus in the mouse. *J Neurosci* **29**:7815–9.

Petrof I, Sherman SM (2013). Functional significance of synaptic terminal size in glutamatergic sensory pathways in thalamus and cortex. *J Physiol* **591**:3125–31.

Pfeiffer R, Bongard J (2007). *How the Body Shapes the Way We Think: A New View of Intelligence*. Cambridge, MA: MIT Press.

Philipp R, Hoffmann KP (2014). Arm movements induced by electrical microstimulation in the superior colliculus of the macaque monkey. *J Neurosci* **34**:3350–63.

Planck M (1949). *Scientific Autobiography and Other Papers* (Gaynor F, Trans). New York: Philosophical Library.

Pogrebkova AV (1970). [Participation of the anterior division of the limbic system in the regulation of respiration]. *Izv Akad Nauk SSSR Biol* **1**:100–3.

Popper K (1959). *The Logic of Scientific Discovery*. London: Hutchinson.

Popper K, Eccles JC (1977). *The Self and its Brain*. Berlin: Springer International.

Raastad M, Shepherd GM (2003). Single-axon action potentials in the rat hippocampal cortex. *J Physiol* **548**:745–52.

Rácz E, Bácskai T, Szabo G, Székely G, Matesz C (2008). Organization of last-order premotor interneurons related to the protraction of tongue in the frog, Rana esculenta. *Brain Res* **1187**:111–15.

Rall W, Shepherd GM, Reese TS, Brightman MW (1966). Dendrodendritic synaptic pathway for inhibition in the olfactory bulb. *Exp Neurol* **14**:44–56.

Ralston HJ 3rd (1971). Evidence for presynaptic dendrites and a proposal for their mechanism of action. *Nature* **230**:585–7.

Ralston HJ 3rd, Herman MM (1969). The fine structure of neurons and synapses in ventrobasal thalamus of the cat. *Brain Res* **14**:77–97.

Ramón y Cajal S (1954). *Neuron Theory or Reticular Theory? Objective Evidence of the Anatomical Unity of Nerve Cells* (Purkiss MU, Fox CA, Trans). Madrid: Consejo Superior de Investigaciones Científicas Instituto Ramon y Cajal. [Originally published in 1933.]

Ramón y Cajal S (1955). *Histologie du Système Nerveux de l'Homme et des Vertébrés* (L. Azoulay, Trans). Madrid: Consejo Superior de Investigaciones Científicas. [Originally published in 1909–1911.]

Ranson SW (1936). Some functions of the hypothalamus: Harvey Lecture, December 17. *Bull N Y Acad Med* **13**:241–71.

Reeves AJ, Amano K, Foster DH (2008). Color constancy: phenomenal or projective? *Percept Psychophys* **70**:219–28.

Riggs LA (1976). Human vision: some objective explorations. *Am Psychol* **31**:125–34.

Robertson B, Kardamakis A, Capantini L, Pérez-Fernández J, Suryanarayana SM, Wallén P, Stephenson-Jones M, Grillner S (2014). The lamprey blueprint of the mammalian nervous system. *Prog Brain Res* 2014; **212**:337–49.

Robson JA, Hall WC (1977a). The organization of the pulvinar in the grey squirrel (Sciurus carolinensis). I. Cytoarchitecture and connections. *J Comp Neurol* **173**:355–88.

Robson JA, Hall WC (1977b). The organization of the pulvinar in the grey squirrel (Sciurus carolinensis). II. Synaptic organization and comparisons with the dorsal lateral geniculate nucleus. *J Comp Neurol* **173**:389–416.

Rockland KS (1996). Two types of corticopulvinar terminations: round (type 2) and elongate (type 1). *J Comp Neurol* **368**:57–87.

Rockland KS (1998). Convergence and branching patterns of round, type 2 corticopulvinar axons. *J Comp Neurol* **390**:515–36.

Sahraie A, Trevethan CT, MacLeod MJ, Urquhart J, Weiskrantz L (2013). Pupil response as a predictor of blindsight in hemianopia. *Proc Natl Acad Sci USA* **110**:18333–8.

Sakmann B (2017). From single cells and single columns to cortical networks dendritic excitability, coincidence detection and synaptic transmission in brain slices and brains. *Exp Physiol* 102:489–521.

Schiller PH, Kendall GL, Slocum WM, Tehovnik EJ (2008). Conditions that alter saccadic eye movement latencies and affect target choice to visual stimuli and to electrical stimulation of area V1 in the monkey. *Vis Neurosci* 25:661–73.

Schneider G (1969). Two visual systems. *Science* 163:895–902.

Schwartz ML, Dekker JJ, Goldman-Rakic PS (1991). Dual mode of corticothalamic synaptic terminals in the mediodorsal nucleus of the rhesus monkey. *J Comp Neurol* 309:289–304.

Scoville WB, Milner B (1957). Loss of recent memory after bilateral hippocampal lesions. *J Neuropsychiatry Clin Neurosci* 12:103–13.

Shepherd GM (1991). *Foundations of the Neuron Doctrine.* Oxford: Oxford University Press.

Shergill SS, Samson G, Bays PM, Frith CD, Wolpert DM (2005). Evidence for sensory prediction deficits in schizophrenia. *Am J Psychiatry* 162:2384–6.

Sherman D, Fuller PM, Marcus J, Yu J, Zhang P, Chamberlin NL, Saper CB, Lu J (2015). Anatomical location of the mesencephalic locomotor region and its possible role in locomotion, posture, cataplexy, and Parkinsonism. *Front Neurol* 6:140.

Sherman SM (2014). The function of metabotropic glutamate receptors in thalamus and cortex. *Neuroscientist* 20:136–49.

Sherman SM (2016). Thalamus plays a central role in ongoing cortical functioning. *Nat Neurosci* 19:533–41.

Sherman SM, Guillery RW (1998). On the actions that one nerve cell can have on another: distinguishing 'drivers' from 'modulators'. *Proc Natl Acad Sci USA* 95:7121–6.

Sherman SM, Guillery RW (2001). *Exploring the Thalamus.* San Diego, CA: Academic Press.

Sherman SM, Guillery RW (2006). *Exploring the Thalamus and its Role in Cortical Function.* Cambridge, MA: MIT Press.

Sherman SM, Guillery RW (2013). *Functional Connections of Cortical Areas: A New View from the Thalamus.* Cambridge, MA: MIT Press.

Sherrington C (1906). *The Integrative Action of the Nervous System.* New Haven, CT: Yale University Press.

Simms ML, Kemper TL, Timbie CM, Bauman ML, Blatt GJ (2009). The anterior cingulate cortex in autism: heterogeneity of qualitative and quantitative cytoarchitectonic features suggests possible subgroups. *Acta Neuropathol* 118:673–84.

Smith GD, Sherman SM (2002). Detectability of excitatory versus inhibitory drive in an integrate-and-fire-or-burst thalamocortical relay neuron model. *J Neurosci* 22:10242–56.

Sommer MA, Wurtz RH (2008). Brain circuits for the internal monitoring of movements. *Ann Rev Neurosci* 31:317–38.

Somogyi G, Hajdu F, Tömböl T (1978). Ultrastructure of the anterior ventral and anterior medial nuclei of the cat thalamus. *Exp Brain Res* 31:417–31.

Sotnichenko TS (1976). Convergence of the descending pathways of motor, visual and limbic cortex in the cat di- and mesencephalon. *Brain Res* 116:401–15.

Sperry RW (1950). Neural basis of the spontaneous optokinetic response produced by visual inversion. *J Comp Neurol* 43:482–9.

Spillane JD (1981). *The Doctrine of the Nerves: Chapters in the History of Neurology.* Oxford: Oxford University Press.

Sprague JM (1966). Interaction of cortex and superior colliculus in mediation of visually guided behavior in the cat. *Science* 153:1544–7.

Squire LR (1992). Memory and the hippocampus: a synthesis from findings with rats, monkeys and humans. *Psychol Rev* 99:195–231.

Steenland HW, Li XY, Zhuo MJ (2012). Predicting aversive events and terminating fear in the mouse anterior cingulate cortex during trace fear conditioning. *Neurosci* 32:1082–95.

Stein BE, Arigbede MO (1972). Unimodal and multimodal response properties of neurons in the cat's superior colliculus. *Exp Neurol* 36:179–96.

Stein BE, Gaither NS (1981). Sensory representation in reptilian optic tectum: some comparisons with mammals. *J Comp Neurol* 202:69–87.

Steinbach SE (1998). How frivolous litigation threatens good science. *Chron High Educ* 4 Dec:A56.

Stevens CF (1994). What form should a cortical theory take? In: Koch C and Davis JL (Eds), *Large Scale Neuronal Theories of the Brain*, pp 239–55. Cambridge, MA: MIT Press.

Strausfeld NJ, Hirth F (2013). Deep homology of arthropod central complex and vertebrate basal ganglia. *Science* 340:157–61.

Strawson G (1997). The self. *J Consciousness* 4:405–28.

Szentágothai J (1963). The structure of the synapse in the lateral geniculate nucleus. *Acta Anat Basel* 55:166–85.

Talan MI (1966). The influence of the anterior limbic cortex on the contractile function of the urinary bladder in cats. *Dokl Akad Nauk SSSR* 166:1248–51.

Taube JS (1995). Head direction cells recorded in the anterior thalamic nuclei of freely moving rats. *J Neurosci* 15:70–86.

Taube JS (2007). The head direction signal: origins and sensory-motor integration. *Annual Rev Neurosci* 30:181–207.

Taube JS, Goodridge JP, Golob EJ, Dudchenko PA, Stackman RW (1996). Processing the head direction cell signal: a review and commentary. *Brain Res Bull* 40:477–84.

Tehovnik EJ, Slocum WM, Schiller PH (2005). Delaying visually guided saccades by microstimulation of macaque V1: spatial properties of delay fields. *Eur J Neurosci* 22:2635–43.

Teufel C, Kingdon A, Ingram JN, Wolpert DM, Fletcher PC (2010). Deficits in sensory prediction are related to delusional ideation in healthy individuals. *Neuropsychologia* 48:4169–72.

Theyel BB, Llano DA, Sherman SM (2010). The corticothalamocortical circuit drives higher-order cortex in the mouse. *Nat Neurosci* 13:84–8.

Thompson E, Varela FJ (2001). Radical embodiment: neural dynamics and consciousness. *Trends Cogn Sci* 5:418–25.

Torrigoe Y, Blanks RH, Precht W (1986). Anatomical studies on the nucleus reticularis tegmenti pontis in the pigmented rat. II. Subcortical afferents demonstrated by the retrograde transport of horseradish peroxidase. *J Comp Neurol* 243:88–105.

Van Essen DC, Anderson CH (1990). Information processing strategies and pathways in the primate retina and visual cortex. In: Zornetzer SF, Davis JF, Lau C (Eds), *Introduction to Neural and Electronic Networks*, pp 43–72. Orlando, FL: Academic Press.

Van Essen DC, Anderson CH, Felleman DJ (1992). Information processing in the primate visual system: an integrated systems perspective. *Science* 255:419–23.

Van Horn SC, Sherman SM (2007). Fewer driver synapses in higher order than in first order thalamic relays. *Neuroscience* 475:406–15.

Van Horn SC, Erişir A, Sherman SM (2000). Relative distribution of synapses in the A-laminae of the lateral geniculate nucleus of the cat. *J Comp Neurol* 416:509–20.

Vann SJ (2013). Dismantling the Papez circuit for memory in rats. *Elife* 2:e00736.

Varela FJ (1996). Neurophenomenology: a methodological remedy for the hard problem. *J Conscious Stud* 3:330–49.

Varela FJ, Thompson E, Rosch E (1993). *The Embodied Mind: Cognitive Science and Human Experience*. Cambridge, MA: MIT Press.

Veinante P, Lavallée P, Deschênes M (2000). Corticothalamic projections from layer 5 of the vibrissal barrel cortex of the rat. *J Comp Neurol* 424:197–204.

von Holst E, Mittelstaedt H (1950). The reafference principle. Interaction between the central nervous system and the periphery. In: Martin R (Ed and Trans), *Selected papers of Erich von Holst: The Behavioural Physiology of Animals and Man*, pp 139–73. Coral Gables FL: University of Miami Press.

Vukadinovic Z (2011). Sleep abnormalities in schizophrenia may suggest impaired trans-thalamic cortico-cortical communication: towards a dynamic model of the illness. *Eur J Neurosci* 34:1031–9.

Vukadinovic Z (2014). NMDA receptor hypofunction and the thalamus in schizophrenia. *Physiol Behav* 131:156–9.

Vukadinovic Z (2015). Elevated striatal dopamine attenuates nigrothalamic inputs and impairs transthalamic cortico-cortical communication in schizophrenia: a hypothesis. *Med Hypotheses* 84:47–52.

Vukadinovic Z, Rosenzweig I (2012). Abnormalities in thalamic neurophysiology in schizophrenia: could psychosis be a result of potassium channel dysfunction? *Neurosci Biobehav Rev* 36:960–8.

Walker AE (1938). *The Primate Thalamus*. Chicago, IL: University of Chicago Press.

Wallace SF, Rosenquist AC, Sprague JM (1990). Ibotenic acid lesions of the lateral substantia nigra restore visual orientation behavior in the hemianopic cat. *J Comp Neurol* 296:222–52.

Walls GL (1953). The lateral geniculate nucleus and visual histophysiology. *Uni Calif Publ Physiol* 1:1–100.

Warren RM, Warren RP (1968). *Helmholtz on Perception: Its Physiology and Development*. New York: Wiley.

Weiskrantz L, Warrington EK, Sanders MD, Marshall J (1974). Visual capacity in the hemianopic field following a restricted occipital ablation. *Brain* 97:709–28.

Westheimer G (2008). Was Helmholtz a Bayesian? A review. *Perception* 37:1–11.

Wheeler M (2007). *Reconstructing the Cognitive World: The Next Step*. Cambridge, MA: MIT Press.

Whittle S, Chanen AM, Fornito A, McGorry PD, Pantelis C, Yücel M (2009). Anterior cingulate volume in adolescents with first-presentation borderline personality disorder. *Psychiatry Res* 172:155–60.

Wickelgren BG, Sterling P (1969). Effect on the superior colliculus of cortical removal in visually deprived cats. *Nature* 224:1032–3.

Wilson JR, Hendrickson AE (1981). Neuronal and synaptic structure of the dorsal lateral geniculate nucleus in normal and monocularly deprived Macaca monkeys. *J Comp Neurol* 197:517–39.

Witt I, Hensel H (1959). Afferente Impulse aus der Extremitätenhaut der Katze bei thermischer und mechanischer Reizung. *Pflügers Archiv* 268:582–96.

Wolpert DM, Flanagan JR (2001). Motor prediction. *Curr Biol* 11:R729–R732.

Wolpert DM, Miall RC (1996). Forward models for physiological motor control. *Neural Netw* 9:1265–79.

Wright MJ, Bishop DT, Jackson RC, Abernethy B (2013). Brain regions concerned with the identification of deceptive soccer moves by higher-skilled and lower-skilled players. *Front Hum Neurosci* 7:851.

Wurtz RH (2008). Neuronal mechanisms of visual stability. *Vision Res* 48:2070–89.

Wykoff RW, Young JZ (1956). The motorneuron surface. *Proc R Soc Lond B Biol Sci* 144:440–50.

Young JZ (1936). The giant fibres and epistellar bodies of cephalopods. *Q J Micr Sci* 78:367–86.

Young JZ (1939). Fused neurons and synaptic contacts in the giant nerve fibres of cephalopods. *Phil Trans R Soc B* 229:465–503.

Young JZ (1950). *The Life of Vertebrates*. Oxford: Clarendon Press.

Yu L, Xu J, Rowland BA, Stein BE (2014). Development of cortical influences on superior colliculus multisensory neurons: effects of dark-rearing. *Eur J Neurosci* 37:1594–601.

Zeki S (1993). *A Vision of the Brain*. Oxford: Blackwell Scientific Publications.

Index

Tables and figures are indicated by an italic *t* and *f* following the page number.